Managing the PSTN Transformation

A Blueprint for a Successful Migration to IP-Based Networks

Managing the PSTN Transformation

A Blueprint for a Successful Migration to IP-Based Networks

Sandra Dornheim

Deutsche Telekom AG

CRC Press
Taylor & Francis Group
Boca Raton London New York

CRC Press is an imprint of the
Taylor & Francis Group, an **informa** business

CRC Press
Taylor & Francis Group
6000 Broken Sound Parkway NW, Suite 300
Boca Raton, FL 33487-2742

First issued in paperback 2020

ISBN 13: 978-0-367-57597-7 (pbk)
ISBN 13: 978-1-4987-0103-7 (hbk)

Visit the Taylor & Francis Web site at
http://www.taylorandfrancis.com

and the CRC Press Web site at
http://www.crcpress.com

Contents

Preamble

Claudia Nemat, Board Member, Europe and Technology, Deutsche Telekom; and Kerstin Günther, Senior Vice President, Technology Europe, Deutsche Telekom.

The IP transformation blueprint described in this book is the product of countless hours of hard work by hundreds of individuals over almost 2 years. It is not just a theoretical cookbook that tells you how this could work. This actually happened—we at Deutsche Telekom migrated the entire public switched telephone network in Macedonia to an Internet protocol-based platform. Our colleagues in Macedonia and everyone else who supported them are real pioneers, within the Deutsche Telekom Group, and within the entire industry. We are publishing our IP transformation blueprint because the lessons we learned and the experience we gained in Macedonia can easily be scaled up and applied in other markets around the world to benefit everyone.

To the untrained ear, it sounds simple: you switch the phone lines from one technology to the other. Nothing could be further from the truth. Our experience has shown that a successful migration relies on key aspects beyond the technological side. Therefore, this work also includes a product portfolio roadmap, a commercial roadmap, an NT/IT roadmap, and several business cases. It is our firm belief that everything we learned during the network migration in Macedonia can be applied to other countries around the globe. It is simply a question of scale, of taking what we did there and scaling it up to larger network environments. In Macedonia, we encountered every type of challenge imaginable, especially in integrating complex systems such as flight control at their airports, alarms for fire and police response, large enterprise customers; the list goes on.

At its core, this is a question of cooperation and collaboration across corporate functions. Projects like this one show what we can do when our people work closely together across borders and functions. It enables us to properly apply our skill and knowledge—regardless of where they are. This close collaboration is inherent in an IP network.

An IP network is a unified, future-oriented system with an unprecedented capacity for the ever-growing demand for bandwidth. It also further strengthens our position in Europe as a technology leader. It brings us closer to our customers, who can now activate new services within hours; to our partners, who can connect their value-added services to our network within weeks; and it brings our local operations closer to each other.

Our ultimate goal is a fully integrated pan-European network where the technology speaks the same language no matter where it is. That language is IP. This network will one day integrate mobile and fixed-line technology and enable a new cloud-based production model. It will eliminate redundancies, increase efficiency, and pave the way for the value-added services and solutions of the future. This blueprint is therefore, the first vital step toward creating a truly pan-European network.

It is just a matter of time. Until then, please read on and learn all about what our IP transformation meant and means to us.

Claudia Nemat   **Kerstin Günther**
Board Member   *Senior Vice President*
Europe and Technology   *Technology Europe*
Deutsche Telekom   *Deutsche Telekom*

The Authors

Sandra Dornheim is project manager at Deutsche Telekom, Europe and Technology, with a focus on go-to-market and launch management. Over the past 2 years she was responsible for the go-to-market work stream within the IP transformation and PSTN migration initiative at Deutsche Telekom. Seven countries were supported and involved in the project and colleagues from these countries provided us with their valuable feedback and insights, which have also contributed to the book. After graduating with a master's degree at the University of Mainz (Germany) in business and economics, Dornheim started working for Deutsche Telekom. She has 11 years of telecommunication experience within different organizations and projects; however, her focus has always been on marketing and sales. She completed her second master's of science degree in strategic marketing leadership at the Henley Business School in the United Kingdom.

Kerstin Groß works at T-Systems as a program director in Corporate Steering, and is responsible for the Transformation Office of T-Systems. After obtaining a master's degree in business administration from the European Business School (EBS), Oestrich-Winkel (Germany), she wrote a doctorate thesis at the Endowed Chair for Corporate Finance and Capital Markets at the EBS. Following her postgraduate work, Groß began her career in consulting at A.T. Kearney; specializing in three areas: aviation, utilities, and telecommunications. As a result, she decided to continue her focus on telecommunications, and worked as a senior project manager at the Center for Strategic Projects at Deutsche Telekom's headquarters (Bonn, Germany) for 3 years. Within that time she was part of the team developing the first IP transformation blueprint responsible for the network and IT blueprint, migration strategy, and framework as well as product migration. After her time at CSP Groß changed to her current position.

Malte Debus works as a senior project manager for the Center for Strategic Projects within the Deutsche Telekom Group. As part of the team developing the first IP transformation blueprint he was responsible for the business case framework. Following his master's degree in business administration from the University of Marburg (Germany), Debus wrote a doctoral thesis on the efficiency of supervisory boards in German stock corporations. He joined Deutsche Telekom in 2009, and has more than 5 years of telecommunication and consulting experience with a strong focus on restructuring and performance improvement projects.

Frank Achmann is vice president at Deutsche Telekom Technical Service, where he is responsible for E2E failure cost reduction. Prior to this position, he worked at the Center for Strategic Programs within the Deutsche Telekom Group, as the project lead developing the first IP transformation blueprint. Upon receiving a master's degree in business and engineering in Cologne (Germany), he joined the automotive industries. As a program manager he was technically and commercially responsible for several multinational programs developing fuel tank systems. After 7 years, Achmann decided to change his focus to telecommunications and worked as a senior project manager at the Center for Strategic Projects at Deutsche Telekom group headquarters.

Julia Hirschle is a project manager at Deutsche Telekom, Europe and Technology, with a focus on strategic commercial projects and processes. For the past 2 years she has been responsible for European-wide business steering processes and commercial projects within the commercial area of Deutsche Telekom's headquarters. Prior to this, she worked as a project manager in the IP transformation and PSTN migration initiative at Deutsche Telekom with a focus on a European-wide best practices approach. Hirschle graduated from the University of Mannheim (Germany) with a degree in business and economics. She has 4 years of telecommunication and project management experience, specifically in restructuring, technology, and innovation projects.

Introduction

Next-generation networking (NGN) describes key architectural evolutions in telecommunication core and access networks that will be developed over the next years. The general idea behind NGN is one single network for all information and services (voice, data, and all types of media, such as video).

The required shift toward standards-based architectures allowing service providers to create multipurpose platforms that share a common infrastructure is called *IP transformation*. It ensures technological leadership forming a new paradigm for telecommunication businesses with higher service quality standards (see Figure I.1).

While there are several enablers supporting the IP transformation as a broadband rollout (xDSL, FTTx) and IP network optimization (e.g., BNG and TeraStream), a basic but challenging requirement is an all-IP infrastructure and thus the decommissioning of PSTN. To reach this goal, customers and products of the PSTN network need to be transferred to the IP network. This process is called *PSTN migration*.

As an enabler for IP transformation, PSTN migration is primarily focused on cost avoidance, network stability, and minimized churn during the migration process. Potential revenue increases from new IP services or value creation for customers and further efficiencies from

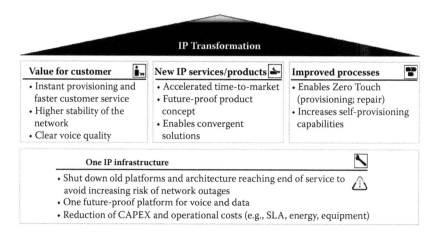

Figure I.1   Benefits of IP transformation.

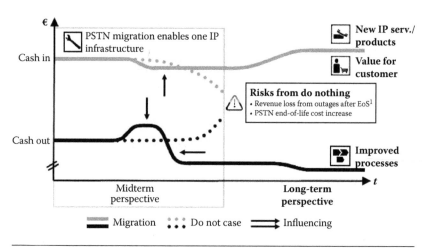

Figure I.2   PSTN migration as an enabler for IP transformation.

process improvements are not within the scope of PSTN migration projects. They need to be addressed separately in a long-term perspective of IP transformation (see Figure I.2).

PSTN migration impacts the whole company for all business functions and with a complete fixed customer base. The largest part of this challenge is posed by the legacy TDM voice platform, the PSTN. All incumbent operators are struggling with their PSTN migration, mostly because they had initially underestimated the complexity and criticality of the issue and as a result:

1. They were not clearly focused on the core purpose of the PSTN migration—which is to avoid costs while minimizing churn—and therefore allowed themselves to be distracted by possible revenue upsides.
2. They tried to approach the problem in a "80/20" fashion, saw some initial success on simple products in simple customer segments, and finally got stuck on the more complex products.
3. They set ambitious top-down targets without prior thorough analysis of the constraints and options at hand, targets which subsequently had to be revised over and over and therefore lost credibility.
4. They framed the issue as decommissioning of a technology, to be led by the chief technology officer (CTO), rather than migration of customers, with heavy involvement and commitment of all functions, particularly the chief marketing officer/chief operating officer (CMO/COO).

During the PSTN migration project within Deutsche Telekom and seven countries, we developed a reference blueprint, summing up the state of the art knowledge on PSTN migrations and key learning.

Here, we provide an overview of this blueprint, sorted by the three key problem areas (Figure I.3):

1. How to capture the financial benefit of the PSTN migration
2. How to realize a 100% complete PSTN migration
3. How to enable and support the PSTN migration

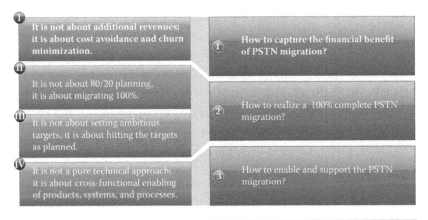

Figure I.3   The PSTN migration challenges, grouped into three key problem areas.

Minimize costs		Avoid churn

Minimize costs

- **Avoid unscheduled investment** to enable migration and related costs
- **Resource need** for migration/after migration support is **underestimated**
- **Unclear migration strategy** limits transparency on migration cost and its optimization

Maximize financial benefit

Avoid churn

- **Churn** needs to be considered carefully especially **during forced migration**
- Impact of PSTN migration on churn is **not transparent**
- PSTN migration is not about 80/20 planning, it is about **migration** of **100% of customers** while avoiding churn

Optimize time

- Top-down **targets need to be verified** and aligned by all involved functions
- Migration **timing** is impacted by **several interdependent factors**, making planning difficult
- **Delay risk is high** with drastic impact on cash line

Exhaustive migration plan required to ensure 100% customer migration with maximal financial benefit

Figure I.4   Triple constraints of PSTN migration.

How to Capture the Financial Benefit of PSTN Migration

As a key requirement for IP transformation, PSTN migration is a necessity which has to be realized in a financially optimal way considering both costs and potential revenue loss. Thus, it is not about additional revenues, it is about cost avoidance and churn minimization.

The financial benefit of PSTN migration is impacted by three dimensions—costs, churn, and time. The management of each dimension and their interaction is highly complex and provides challenges in every migration project (see Figure I.4). All three dimensions need to be reflected in the business case model.

The business case can be simply constructed by comparing migration costs to cost savings due to the shutdown of PSTN platforms. Experience shows migration from PSTN to IP technology costs around 30€ ($38) to 60€ ($75) per subscriber, distributed over 3 to 5 years, while the expected cost benefits from switching off the PSTN platform are around 10€ ($13) per subscriber per year.

However, these figures only provide a rough orientation and it has to be kept in mind that

- It is often unclear which costs are migration related and thus within scope.

- Migration costs are heavily influenced by required resources for each customer segment, especially toward the end of cleansing an area.
- Savings of PSTN migration may be underestimated, because they do not account for an avoidance in cost explosion or churn if some of the switches actually do reach the end of their useful lifetime.
- Business case results are highly sensitive to migration-related churn; already 2% of total customer base churn can alienate the financial benefit of the business case.

Hence, the business case has to be extended, taking various additional factors into account.

Additionally, for decision-relevant business cases, the question—what to compare PSTN migration with—is not an easy one. Comparing it to an imaginary "flat line" do-nothing scenario is not realistic, as do nothing yields a strong rise in costs, as well as a high risk of network failures and a total failure of the PSTN in the long run. Thus, it is not a real option and the comparison of the PSTN migration scenario should be conducted against a realistic scenario or minimal change—usually a midterm interim solution such as prolonged PSTN usage or soft switches.

Hence, the benefit of PSTN migration needs to be calculated by comparing incremental financial implications of the specific interim and full migration options of the countries with regard to revenue, operational cost development, and CAPEX (capital expenditure) investment. Having transparency in these options as well as parallel enabler projects (e.g., investments into higher broadband coverage) are key challenges and prerequisites for solid and non-overlapping business case calculations.

The financial benefit of a full migration scenario versus interim solution can be influenced by four major drivers:

- *Minimize churn*: A customer facing problems, while being migrated or later using IP services, will always churn with his full contract revenue from voice, broadband, TV, and other services.
- *Migration costs*: The level of many expenditures, such as IMS or access network investment, is determined by the technical starting point of the country; but resource needs for customer

migration and levers for optimization (e.g., remote migration) should be carefully considered.

- *Migration timeline*: Migration of the full customer base should always be accomplished in an ambitious timeframe to minimize the periods with negative cash flow.
- *Long-term savings*: The main expected benefits of switching off the PSTN platform are savings in energy costs, service level agreements, and personnel costs in planning, assembling, and maintenance.

To support each country with this challenge, we have developed a standardized business case framework. Besides overarching benefit calculations this helps to

- Provide a solid and comparable basis to orchestrate CAPEX, OPEX (operational expenditure), and revenue discussion within and across all functions (e.g., marketing, sales, customer care, NT, IT, regulatory).
- Ensure cross-functional alignment for target setting during the conception phase of the project and rollout preparation as well as final execution of the migration plan.
- Faster align the country business case results with shareholder Deutsche Telekom AG through a common understanding of the impact from PSTN migration.

How to Realize a 100% Complete PSTN Migration?

One key challenge of PSTN migration is that it is not about 80/20 planning, it is about migrating 100%. This holds true not only for the total project and the PSTN shutdown but also for the individual area shutdown. However, these area shutdowns are the main driver of savings. Thus, the question is not, how fast can you migrate a large number of customers, but how long does it take to migrate the last customer. It also shows the migration status in 2012, which had already migrated a large number of customers, but still has left the more complex business and wholesale customers, who might block the shutdown of the focus areas.

To mitigate the resulting risk, careful planning has to be done, which is not about setting ambitious targets; it is about hitting the targets as planned.

Consequently, migration planning needs to realistically detail how to contact and migrate customers, and also when to migrate each customer. To realize this and optimize the trade-off between costs, time, and churn, some key learning is applied in the migration planning framework presented in the blueprint:

- Migrate by area to maximize savings, migrate by product to leverage spare resources (costs and time).
- Start early with customized solutions and wholesale solutions, as they often block the clearance of areas (time).
- Maximize leverage of natural migration (NM), as it is the most optimal resource and most customer-friendly way for migration (costs).
- Manage all types of resources choosing a trade-off on churn, cost, and/or time, and check readiness to make those sacrifices (costs, time, and churn).

To ensure the implementation of learning and a smooth migration process, the framework focuses on the two major resource constraints of product development and migration resources. The resulting migration plan comprises not only the sequencing of product development but also the planning of the migration itself, defining how each customer is migrated (migration strategy), and when (migration sequence).

The planning process starts with the optimization of product development, defining not only the development start date but also the date of substitute readiness. The prioritization of products follows two imperatives:

- High-volume mass products first as their early readiness increases natural migration and resource leverage.
- Complex products first as they take a long time to develop, migrate, and form the major roadblocks.

The migration sequencing approach strives to maximize savings while managing available migration resources and reducing the delay risks of the project. As savings are mainly driven by the shutdown of an area, migration should be done in an area-by-area approach. However, to leverage spare resources before and during an area migration, a migration by product is used as well.

Identifying the product sequence within the product-by-product approach, a framework is provided by mainly applying two recommendations:

- *Complex products*: Start with multiarea products to reduce roadblocks for area cleansing, start with complex products to mitigate long migration duration.
- *Standard products*: Focus on mass products with expected limited success of natural migration.

Area sequencing forms the most complicated task, as many resource and timing constraints are implied on the area-by-area migration combined with strong interdependencies of sequencing and resource requirements. To break this circularity issue, the sequencing is done in a scenario approach taking timing constraints, savings on migration costs, and early migration savings into account. The scenarios are combined and optimized by different levers to ensure sufficient resource availability on different dimensions (CAPEX/OPEX, customer care FTEs, regional technical services FTEs).

How to Enable and Support the PSTN Migration?

In preparation of PSTN migration three major adjustments are required, not only in NT and IT, but also in the product portfolio and processes, which can impact various functions. Hence, PSTN migration is not a pure technical approach; it is about a cross-functional enabling of products, systems, and processes:

- The PSTN product portfolio needs to be transferred into the IP world and thus the new value-cost-optimized IP product portfolio needs to be created.
- NT/IT need to be enabled for IP and the migration demands vast adjustments in systems.
- IT processes for migration need to be defined and the majority of existing processes adapted.

The development of the new IP product portfolio is complex and crucial due to the large scale covering all PSTN product substitutes and due to the strong regulatory involvement throughout the whole process. Both aspects impose not only high effort and huge project risks but also have a major impact on future positioning and revenues. Thus, an approach must be used ensuring a carefully cost-benefit-optimized and approved product portfolio.

As a basis for all consideration, transparency regarding the current PSTN product portfolio and its features, as well as the existing options for the new portfolio, have to be generated allowing the deduction of required actions per feature.

One of these actions is a cost-benefit analysis of the possible features to be invested in. The blueprint provides an evaluation framework assessing feature value and costs resulting in the list of features to be implemented in the IP world and identification of features to be retired.

Building on the IP feature portfolio, the new product portfolio needs to be created and optimized along with various criteria, such as financial advantages and customer experience. After defining each product, the approval by the regulator is required for resulting adjustments and documents (e.g., terms and conditions, reference offers, pricing), which for best practices and argumentation lines are provided in the blueprint.

Furthermore, the selected product portfolio needs to be technically developed and implemented as well as launched. In addition to the definition of the new product portfolio, NT and IT infrastructure needs to be adjusted to prepare for the IP world and the migration itself. This requires careful planning as large parts of the network are impacted creating high effort and high risks. Also, process adjustments are characterized by large-scale, high business risks, and a strong impact on migration success.

Book Structure

In general the book is divided into six chapters, including five functional work streams embedded in an overarching work stream. The main topics of these work streams, which will be presented in the following chapters, are:

1. "Overarching Topics" provide an overview of major organizational overarching topics, for example, the project structure with roles and responsibilities, the PSTN migration process,

as well as a checklist of the main questions and risks that should be considered during the project.

2. "Migration Plan" within a three-step approach including prioritization logic to ensure efficient rollout planning.

3. "Product Portfolio Roadmap" describes the necessary processes and steps for transferring the PSTN products to an IP world. This includes both the analysis of the features and the product definition.

4. "NT/IT Roadmap" consists of three substreams: network target architecture and technical product development; NT rollout—adapting the current network structure to enable PSTN migration including rollout planning and technical training; and IT roadmap—adapting current systems to enable new IP features/products.

5. "Business Case Framework" defines a common business case framework enabling cross-functional steering during conception, rollout preparation, and execution.

6. "Go to Market" ensures a value-based migration of the full customer base with an efficient rollout plan as well as a detailed plan for marketing introduction, a sales channel plan, and management of the resource needs during migration.

List of Abbreviations

3GPP: 3rd Generation Partnership Project
AbA: Area-by-area
ADSL: Asymmetric digital subscriber line
ARPU: Average revenue per user
AS: Application server
ATL: Above the line
ATM: Asynchronous transfer mode
BAU: Business as usual
BB: Broadband
BC: Business customer
BNG: Broadband network gateway
BSA: Bitstream access
BSS: Business subsystem
BTL: Below the line
CC: Customer care
CDR: Call detail record
CLF: Customer locations function
CO: Central office
CPE: Customer premises equipment
CPS: Carrier pre-selection
CT: Crnogorski Telekom
CRM: Customer relationship management

CS: Carrier selection
DHCP: Dynamic host configuration protocol
DLU: Digital line unit
DSL: Digital subscriber line
DSLAM: Digital subscriber line access multiplexer
DTMF: Dual-tone multi-frequency
DWH/BI: Data warehouse/business intelligence
E2E: End-to-End
EC: Emergency call
EoS: End of service
ETL: Extract, transform, load
ETSI: European Telecommunications Standards Institute
F2F: Face-to-face
FM: Forced migration
FMS: Fault management system
FTE: Full-time equivalent
FTTH: Fiber to the home
FTTx: Fiber to the x
FW: Firmware
GPON: Gigabit passive optical network
GUI: Graphical user interface
HR: Human resources
HT: Hrvatski Telekom
HW: Hardware
iAD: Integrated access device
IC: Interconnection
IMS: IP multimedia subsystem
ISDN: Integrated services digital network
ITU: International Telecommunication Union
IVR: Interactive voice response
LE: Local exchange
MGC: Media gateway controller
MGW: Media gateway
MKT: Makedonski Telekom
MMTel: Multimedia telephony
MSAN: Multi-service access node
MT: Magyar Telekom
MTBF: Mean time between failure

NGN: Next-generation network

NM: Natural migration

NPV: Net present value

NRA: National regulatory agency

OLT: Optical line terminal

OSS: Operating subsystem

PbP: Product-by-product

PBX: Private branch exchange

POS: Point of sales

POTS: Plain old telephone service

PSTN: Public switched telephone network

QoS: Quality of service

RAS: Revenue assurance system

RC: Residential customer

RO: Reference offer

RSU: Remote subscriber unit

SDH: Synchronous digital hierarchy

SIP: Session initiation protocol

SLA: Service level agreement

SSW: Soft switch

ST: Slovak Telekom

SW: Software

T&C: Terms & conditions

TDM: Time division multiplexing

TISPAN: Telecoms & Internet converged Services & Protocols for Advanced Networks

TPD: Technical product development

TS: Technical service

ULL: Unbundled local loop

USO: Universal service obligations

VoBB: Voice over broadband

VSE: Very small enterprise

WACC: Weighted average cost of capital

WAN: Wireless access network

WLR: Wholesale line rental

WS: Wholesale

WWMS: Workflow and workforce management system

xDSL: X digital subscriber line

Project Team Members for IP Transformation

Sven Hischke
Deutsche Telekom

Thorsten Albers
Makedonski Telekom

Bozidar Poldrugac
Hrvatski Telekom

Chapter 1: Overarching Topics

Julia Hirschle
Deutsche Telekom

Liljana Najdenova
Makedonski Telekom

Frank Achmann
Deutsche Telekom

Vladimir Cosic
Hrvatski Telekom

Alexandra Dierkes
Deutsche Telekom

Sandra Dornheim
Deutsche Telekom

Felix Früh
Deutsche Telekom

Peter Grünthal
Deutsche Telekom

Marc Vollweiler
Deutsche Telekom

Chapter 2: Migration Plan

Kerstin Groß
Deutsche Telekom

Sandra Dornheim
Deutsche Telekom

Frank Achmann
Deutsche Telekom

Wolfgang Hauptman
Deutsche Telekom

Malte Debus
Deutsche Telekom

Chapter 3: Product Portfolio Roadmap

Kerstin Groß
Deutsche Telekom

Peco Nedelkovski
Makedonski Telekom

Dario Winkler
Hrvatski Telekom

Kai Schmidt
Deutsche Telekom

Aleksandar Chekorov
Makedonski Telekom

Askan Schmeißer
Deutsche Telekom

Vladimir Cosic
Hrvatski Telekom

Chapter 4: NT/IT Roadmap

Kerstin Groß
Deutsche Telekom

Nikolaus Gieschen
Deutsche Telekom

Stefan Andres
Deutsche Telekom

Laszlo Hobinka
Deutsche Telekom

Dragan Corbeski
Makedonski Telekom

Slobodanka Nikolovska
Makedonski Telekom

Dejan Dabevski
Makedonski Telekom

Natasa Samardziska
Makedonski Telekom

Stevco Dimovski
Makedonski Telekom

Snezhana Tilovska
Makedonski Telekom

Chapter 5: Business Case Framework

Malte Debus
Deutsche Telekom

Anja Fischer
Deutsche Telekom

Boris Batelic
Hrvatski Telekom

Liljana Najdenova
Makedonski Telekom

Elena Cvetkovska
Makedonski Telekom

Chapter 6: Go to Market

Sandra Dornheim
Deutsche Telekom

Miljenko Karakas
Hrvatski Telekom

Irena Lokvenec
Makedonski Telekom

Moritz Kriegsmann
Deutsche Telekom

Aleksandra Janiszewska
Deutsche Telekom

Valerie Pansini
Deutsche Telekom

Hinko Kalpic
Hrvatski Telekom

Andrea Szirtes
Magyar Telekom

1

OVERARCHING TOPICS

The main topics of the overarching work stream include the following:

1. The PSTN migration process and interdependencies give an overview of IP transformation project steps, main milestones, the timeline, and show the interdependencies between all streams.
2. Project organization describes the project organization structure, roles and responsibilities of each stream during different phases of the project, key meetings, as well as escalation and decision processes.
3. Environmental status quo lists the main elements and questions which require attention from the external and internal points of view.
4. Scenario analysis assesses the different migration scenarios with a qualitative assessment of scenarios.
5. A list of standard key performance indicators (KPIs) provides mandatory and optional KPIs that should be considered during the PSTN migration process.
6. The section on risks is dedicated to project risks, their multiple dimensions, and impacts. As an output from this section, the risk list with possible measures for mitigations is developed.
7. End-to-End (E2E) process adjustments provide examples of processes that should be changed during the migration as well as new processes that could or should be introduced.

1.1 The PSTN Migration Process and Interdependencies

The overall PSTN migration process shows the major process steps and the interaction between work streams. First, it should provide a high-level view of the total process duration; and second, it should point out main interdependencies between the streams. These interdependencies can be due to input/output relations between the streams as well as in terms of milestones to be provided to ensure the overall project success. The key milestones can be used as scheduled checkpoints for main management decisions.

The overall PSTN migration process is an overview of all major project steps describing what to do during PSTN migration (see Figure 1.1).

The PSTN migration is divided into three general steps: "conception," "rollout preparation," and "execution."

Within the overall process four main project gates have to be passed:

- *Gate I*: The first board of management (BoM) decision confirms the general project start and the provision of a first budget indication.
- *Gate II*: After approximately 6 months, it provides the first insight on initial migration plan, technical target picture, and financial impact (top-down aspiration, for more details see business case work stream, in Section 5.2.1, Chapter 5). The main goal is to get the commitment of involved functions to stick with basic input and assumptions.
- *Gate III*: After the finalization of the product feature portfolio and the final migration plan, the business case will be validated based on a bottom-up calculation. These results are the base for the BoM decision of the rollout start (gate III). This gate provides transparency on the benefit of PSTN migration before major investment is done. In addition, the first drafts of functional rollout plans are ready and the main goal is to secure commitment of involved functions for rollout.
- *Gate IV*: The last gate after the rollout preparation of all work streams is the BoM decision of the migration start.

The overall timeline is shown in the PSTN migration process and the approximate duration from project start to migration start is 20 months. It is only an illustrative project process; duration differs between countries, depending, for example, on:

- Network status, broadband coverage, and penetration prior to migration start
- Size of customer base that should be migrated
- Existing product portfolio and regulatory requirements

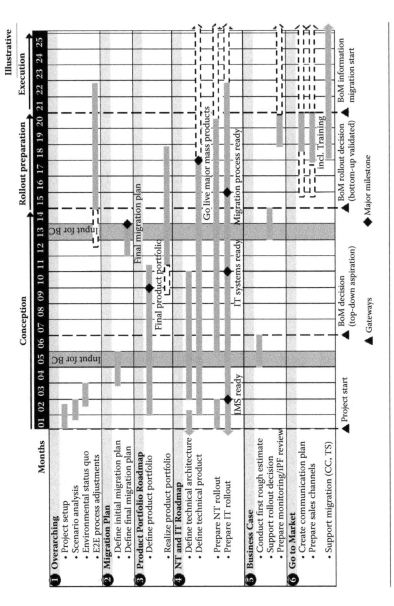

Figure 1.1 Overall PSTN migration process.

- Management expectation for finalization of the project
- Handling of problems and risks during the execution
- Available resources (e.g., FTE, CAPEX)

During the execution phase, there should be a continuous monitoring of defined KPIs, updates of the business case in case of bigger changes of assumptions, and alignment.

1.2 Project Organization

PSTN migration requires an efficient project structure, involving a cross-functional and highly integrated project team. Roles and responsibilities inside the initiative are established as well as the governance structure and an escalation model. Due to its complexity, it needs absolute commitment from all areas in the company and support from top management.

1.2.1 Project Structure: Roles and Responsibilities

IP transformation is a central initiative of Deutsche Telekom (Europe). Within that initiative, a central team is formed to coordinate within the European countries all IP transformation topics referring to PSTN migration as well as BNG and TeraStream. Accordingly, the initiative organization is divided into PSTN migration with its functional and business case work streams developed in the blueprint and into two work streams for BNG and TeraStream (both BNG and TeraStream are not part of this book).

Every stream has a lead responsible for the functional topics (product portfolio roadmap, migration plan, go to market, NT/IT roadmap, business case). In addition, the two responsible for the PMO activities are required for coordination of the central team and communication toward the countries. As a common decision and steering platform an IP transformation EU core team and an overall SteerCo team are established (see Figure 1.2).

On a country level, local IP projects are set up to interact with the central team. In this book, the focus is on PSTN migration, omitting further IP transformation topics.

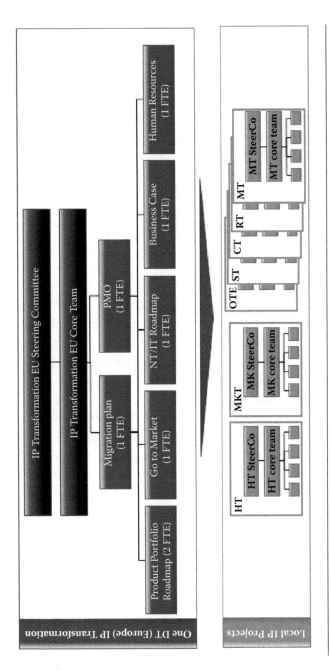

Figure 1.2 Initiative structure.

The experience from the countries have shown that PSTN migration is a complex project that requires aligned involvement of all functions, which has proven to be highly complex and difficult to manage. Thus, all business functions need to collaborate in an efficient way—until all customers are migrated.

Based on this discussion, the involved parties are structured into the functional topics represented by the work streams (overarching topics, product portfolio roadmap, migration plan, NT/IT roadmap, business case, go to market) leading to a cross-functional project organization. Human resources must be integrated from the beginning to support and assure a comprehensive resource planning in the migration plan. The structure is completed by the E2E responsible for the customer segment (business, residential, and wholesale customers) in order to ensure the overall customer perspective.

1.2.2 Escalation

In the first escalation instance within the local IP projects for any risk or issue is the core team. It immediately reviews and assesses the escalation. The evaluation will be recorded in the project management risk register or issue log. Mitigating actions will be taken and tracked by the project management until the issue is resolved or the risk is reduced to an acceptable level.

The next escalation step involves the local steering committee.

The ultimate point of escalation is to the IP transformation steering committee, but this must pass through the escalation channels: project/stream leader → program management → local core team. Project escalations that are not raised to the project/stream leader may not have the appropriate support to resolve them.

1.3 Environmental Status Quo

In the beginning of the preparation phase, an analysis of the environment and the situation of the company and its competitors is required. Several internal and external factors impact the IP transformation; consequently all possible elements have to be considered and defined through a checklist prior to the start of the migration. However, these factors vary by country and need to be adjusted and complemented by country specifics.

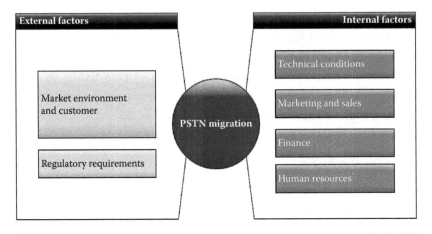

Figure 1.3 External and internal factors.

Environmental factors can be divided into external factors and internal factors. External factors are related to the market environment, customers, and regulatory conditions. Internal factors and questions are related to all areas of operation of the company: technical, marketing, sales, finance, and human resources (see Figure 1.3).

1.3.1 External Factors

The market environment should be carefully monitored because it has several consequences on the PSTN migration process. First, the telecommunication market should be considered in terms of broadband penetration and fixed/mobile penetration as well as development trends. These factors have a tremendous influence on the number of customers that should be migrated over the years, possible natural churn, and the number of broadband customers that should be forecasted. In addition, the customer structure depends on competition movement and the general microeconomic situation, which influence the purchasing power of the population. All these elements have an impact on cost and on the revenue side and should be carefully monitored.

The regulatory environment is one of the main risk drivers for PSTN migration. It can impact the timeline of the project and its financial benefits, for example, by revenue downsizing with possible changes in the interconnection fees or changes in wholesale products. The regulatory requirements can also indirectly impact PSTN migration by affecting other enabler projects, for example, the regulation of fiber in the access could influence the estimated broadband penetration.

A special focus during the migration should be placed on wholesale customers because of the complex processes of negotiation and the complexity of the products. A delay in migration of wholesale customers forms a high risk to the project timeline and its costs.

1.3.2 Internal Factors

Internal factors that should be considered during the PSTN migration are related to technical, marketing/sales, financial topics, and HR/ stakeholder analysis. More details on internal factors are listed hereafter.

From a technical point of view, the main aspects are:

- Status of the network and IT applications, considering both the availability and the readiness before the actual start of the migration
- Provisioning of HR resources for the period of migration
- Alignment of modified processes with all parties involved
- Validation of services through tests prior to the commercial launch
- Follow-up on performance, through regular reports and KPIs

From a marketing point of view, the main questions are related to:

- Alignment of the product portfolio for each customer segment and harmonization with the technical area
- Possibility to develop new products as a source of additional revenue
- Marketing communication, with a strategy and plan for both an internal and external level

Marketing and sales are strongly involved in the PSTN migration process, especially customer care and sales needed to manage the contacts with customers within different channels. This involves both a thorough planning of customer contacts and a follow-up on the performance according to the established KPIs.

The regulatory and financial environment should also be checked prior to migration. Questions for the regulatory area are mainly related to on-time communication with the regulatory agency, wholesale matters, as well as on-time preparation of all the necessary documents, especially the terms and conditions.

Finance questions are related to:

- Input for the business case in terms of cost/savings and revenue potential
- Permanent monitoring of significant changes in estimations
- Permanent monitoring of risk
- Alignment of the PSTN migration project with financial planning cycles and with other enabler projects

Finally, the human resources environment should be checked and a stakeholder analysis should be conducted. Human resources issues can heavily influence the success of the migration. Restrictions regarding the flexibility of resource allocation can occur from several sources such as inflexible agreements with social partners or ongoing headcount reduction programs.

A thorough analysis must be carried out to identify all the relevant stakeholders who will support the migration and those who might slow it down or will try to avoid it. Through this analysis, the visible risks can be prevented and mitigated.

Some general lessons learned in the area of human resources that should be considered for migration planning are:

- PSTN migration requires additional resources (internal/external) during the migration phase and additional OPEX for training, upskilling, and reskilling
- Additional resources are not only needed in the technology area but also in the commercial/customer-related areas
- Significant savings within the HR focus (FTE, OPEX) cannot be expected before the end of the migration
- Several countries have trained their technicians as "universal technicians" (legacy and IP) to have both skill sets available
- Cross-functional project team setup is mandatory to achieve the targets

1.4 Scenario Analysis

Through a scenario analysis, it is possible to evaluate the different alternative states and interim solutions that are available to reach an optimal solution for each country. This can differ according to several variables, including the level of the existing network, market trends, management expectations, and costs that need to be considered. By considering alternative possible outcomes and future developments, an analysis of possible future events is conducted.

PSTN/ISDN is a predominant legacy platform in which huge investments have been made in the past and which currently is still one major fixed revenue source for operators. Due to this, it is logical to focus on the evolution and further development of the voice service architecture.

The current network architecture is complex and consists of different networks and network layers. PSTN/ISDN are predominantly voice networks and are completely separate from data networks. The double operation of networks leads to higher costs in maintenance, operation, and development. Furthermore, PSTN/ISDN have reached their end of service (EoS), which leads to increasing operation costs, limited availability of spare parts, and thus an increasing risk of failure. Transforming this architecture and merging the networks to a standardized and lean IP-based architecture is a challenge. However, the simple and modern technology landscape should lead to decreased costs in maintenance, new possibilities based on convergence of voice, data, and video services, and faster time to market.

In order to identify the optimal way for how to reach the target architecture considering technical and financial impacts, the following scenarios are possible:

0. Do-nothing scenario:

  • No action and investment taken
  • Leading to network failure, resulting revenue loss, and eventually the network breaks down
  • Not considered as a realistic option

1. Interim solutions:

  • Adaptation of current architecture or replacement by substitute version, but no migration to a single network and platform (IMS)
  • Defined as a minimum requirement
  • Only delays PSTN migration

- Two possible options: Prolong PSTN usage with platform upgrades required or implementation of soft switch solution (replacing PSTN network with a separate new one, no customer contract required)

2. One platform (IMS)/full PSTN migration toward NGN IP-based architecture:

- One joint core platform (IMS)
- Full retirement of PSTN with only one network remaining
- Migration of all PSTN customers
- Two possible options for voice-only customers: Migration to MSAN POTS card (emulation of PSTN with no customer impact and contact required) or migration to broadband port (full IP solution, but customer contact and action required)

Combined approaches of interim solutions and full migration at different areas are also possible and form special cases.

A short description of each option and the combined approach is given in the following paragraphs.

1.4.1 Interim Solutions

A PSTN network is a time division multiplexing (TDM)-based hierarchical network existing on international, transit, and local exchanges where the subscribers are connected to remote subscriber units (RSUs) via TDM lines.

Interim solution options for securing the fixed voice service are the following:

1a. Prolonged usage of PSTN network.
1b. Introduction of soft switch (SSW) and media gateway (MGW) elements as a replacement for PSTN exchanges.

Option 1a: Prolonged usage of the PSTN network assumes keeping the TDM-based hierarchical network with international, transit, and local exchanges. In this network, the subscribers are connected to remote subscriber units (RSU) via TDM lines. This is a distributed service architecture, which is from today's point of view, very complex and cost ineffective. It is characterized by a high number of nodes both in core and access part. TDM-based switches are very old, which lead to the following challenges even if usage is prolonged:

- No further production of hardware (processors already out of stock)
- No software support by suppliers (even in the case of SLA extension)
- Source code is not accessible for other suppliers
- No guarantee for functioning exchange by suppliers

Under these circumstances, risk of failure is increased on one side and on the other side there is no possibility for the introduction of additional functionalities (for example, by the regulatory body). Therefore, in order to provide a reliable voice service, this scenario assumes that a change of existing hardware and an upgrade of the current software version of exchanges is necessary, which requires a high volume of CAPEX. The overall network OPEX cost will continue to increase year over year based on increased SLA, energy consumption, and parallel operation of old legacy and IP-based systems.

It should be emphasized that this scenario could be considered only for a limited period of time since vendors will stop development and support of old legacy PSTN equipment.

Option 1b: In comparison to the first scenario, the soft switch option introduces different architectures by decoupling of control (SSW) and transport layer (MGW). Thus, it enables the introduction of a cost effective IP technology in the core transport layer and SIP communication protocol in the control layer. This enables the reduction in the number of core elements where soft switches and media gateway replace several exchanges. TDM transport remains only in the aggregation and access part, so that customers do not need to be touched. However, if the digital line unit (DLU) cannot be used with the soft switches,* full recabling in the central office to the new equipment is necessary, making it costly and time consuming.

As an interim solution, the soft switch solution replaces the old PSTN equipment thus leading to a longer lifetime due to newer soft switches. However, the lifetime of soft switches and with it the possible extension of the voice service based on PSTN (voice services only) are limited, making it only an intermediate solution toward an

---

* DLU reusage is only possible with Siemens switches.

SIP-based network. Consequently, it can be used to buy some time for the time-consuming process of PSTN migration. The soft switch solution also provides the opportunity to replace RSUs with H.248 MSANs, which are capable of all-IP-based fixed voice service.

The downside of this solution is that it is not future proof, since it is only applicable to voice services and it cannot offer multimedia services.

1.4.2 Full Migration

The final IP target picture handles all types of traffic through one platform, the IMS, a basic platform for development of new multimedia customer-oriented services. Its architecture is built from end to end on IP and is characterized by independence of the service layer from the access network. This enables the reuse of common control and service layers for different types of access (copper, FTTH, 2G/3G/LTE, etc.), limiting costly adjustments of those layers. Its access independent characteristic also enables convergent services and service interaction (e.g., missed calls on IP TV screen).

Broadband customers are further serviced via broadband ports and just need to connect their telephone to a VoIP capable integrated access device (iAD), which might have to be provided. Provisioning of voice via broadband line for 2play and 3play customers presents significant savings, since one port is used for all services. However, for voice-only subscribers two options are possible, leading to two different scenarios:

2a. Migration to MSAN POTS cards (emulation of PSTN service).
2b. Migration to broadband port (same solution for all subscribers).

Option 2a assumes the provisioning of voice service for broadband customers via broadband connection (via iAD) and the usage of narrowband MSAN POTS ports for voice-only customers.

Provisioning of voice service via MSAN POTS for voice-only customers allows faster and silent migration (remote provisioning—no in-house work) and reduction of equipment provisioning cost (no iAD is necessary). However, no new services are provided to the customer since it only emulates PSTN voice service. From a strategic point of view, voice as a stand-alone service will have no future, also shown by the permanent declining number of voice-only customers. Thus,

MSAN POTS as a technology emulating voice-only services is an investment in a dying service.

Option 2b assumes full PSTN migration connecting all users via broadband ports to the IMS forming the final target picture:

> As high and sufficient broadband penetration is required, PSTN migration should be planned as an integral component with the broadband access transformation.
>
> Additional costs for this option compared to MSAN POTS are invested for iAD, potential work on customer premises and if the installed broadband capacities are not sufficient, additional investment for broadband ports. Thus, these costs can be higher than the investment in a MSAN POTS card.
>
> The higher costs can be justified in the case of upselling potential during migration as well as in the years to come, as the customer is then required to move to a broadband port anyway.

Consequently, important parameters which have to be taken into account when deciding between two full migration scenarios are:

- Current country's broadband penetration
- Marketing forecast of future broadband penetration
- Likelihood of upselling
- Characteristics of the operator's copper access network (broadband potential depends on the quality of the physical network and the average length of the local loop)

CAPEX in both full migration scenarios is related mainly to IMS, CPE/iAD, MSAN capacities, or broadband ports. Compared with the interim solution, OPEX savings, mainly based on savings in SLA, energy, and personnel costs, are expected. In the long run, it is a future-proof solution that should enable greater efficiencies at lower costs and form a basis for value propositions centered upon applications, services, network access, and data carriage.

1.4.3 *Combined Approach (on an Area Level)*

The different scenarios can also be combined on an area-by-area approach defining which customers of one area will be migrated to all IP, but in other areas remain connected to old legacy platforms.

The connection of customers in nonmigrated areas can be established in three ways:

- Via old PSTN switches, which most likely need to be upgraded
- Via already introduced soft switches
- Via newly introduced soft switches

This is a highly complex solution: It is complex from an operation and maintenance point of view, since different types of equipment are implemented in the network, which also require different resources to operate with. Furthermore, customers cannot always be provided with the same services.

Costs are expected to be higher mainly due to the costs for support contracts and increased personnel cost for operation. Based on these drawbacks, the combined approach should be carefully evaluated.

1.5 Standard Mandatory and Optional KPIs

KPIs represent a useful tool to steer progress in the PSTN migration project and provide full transparency and the achieved results. Data is required from all streams to monitor the process with regard to minimal churn, time, and costs/savings. The standard KPI set focuses on the trade-off between minimal churn, time, and cost/savings.

Laying a base for project steering, the KPIs should be used for:

- Initiating corrective actions and alteration activity timing
- Setting targets, and challenging execution
- Comparing countries and standardized monitoring of the project on a group level

It should be noted that the list of KPIs is not exhaustive and can be extended further based on operator requirements.

1.6 Risk Analysis

Several external and internal risk aspects (parameters) should be considered in the overall project. Each risk has inherent multiple impact dimensions on key business case drivers and project success. Therefore,

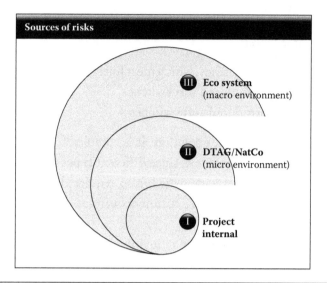

Figure 1.4   Risk sources.

the financial impact should be calculated when these risks can be foreseen and all risks should be reported and followed in order to mitigate them appropriately. The risk quality analysis and assessment is based on the migration plan for the remaining execution phase. Due to the long-term duration of the PSTN migration project, there could be many risk sources during the whole project duration.

During PSTN migration, risks can be inflicted by the project itself, Deutsche Telekom AG or operator level (micro environment), and eco system (macro environment). Therefore, it is important to identify and structure potential sources of risks and define the responsible person by indicating and quantifying the potential impact on churn, costs, and time (Figure 1.4).

The following paragraphs present in detail possible types of risks according to the type of source.

1.6.1 Project Internal Risks

Project internal risks are mainly related to:

- Commitment and involvement of the full organization
- Wrong assumptions of the inputs (e.g., underestimated costs or inputs readiness), wrong estimation for FTE efficiency or competency of resources in the IP world

- Forced migration (FM) process

In particular, forced migration can bear high risks—due to the product complexity (especially in the business segment) and the natural behavior of the customer; several events can occur:

- Churn above defined limits
- Specific substitution products missing (e.g., in business segment)
- Wholesale and business customer migration process is not ready within the planned timeframe

One should be aware that churn due to the PSTN migration (except in cases of poor service quality) is hard to identify. Churn can result from unexpected customer behavior (willingness for migration due to a lack of knowledge) or the need for recontracting the customer ("forcing" customers to look for other offers—especially in the business segment).

Additionally, the impact on the business case in terms of churn, costs, and time is given as an indication. The financial impact should be integrated in the business case, when these risks can be foreseen and have significant impact on the financial results.

1.6.2 Deutsche Telekom AG/NatCo Risks (Micro Environment)

Micro environmental risks could be a result of certain actions of the National Company (NatCo) itself or at the Deutsche Telekom AG group level. These risks are mainly related to:

- Changes in the CAPEX/OPEX such as possible cuts during the year or reprioritizations. These risks could have several impacts:
  - Defined migration timeframe needs to be extended, which could lead to higher OPEX costs.
  - Possible revenue loss/churn due to possible local exchange outages caused by prolonged timeframe.
  - Lower cumulated savings due to prolongation of migration time and time delay of saving achievements. Amount of savings depends on investments made until the risk took place (taking into consideration the level of each LE status).

- Headcount reduction could have an influence on costs, as external resources are needed to replace the reduced headcount. It can also have an impact on the project timeline and thus influence the expected OPEX savings.
- No fulfilled expectations for broadband penetration require a higher incremental CAPEX for PSTN migration and could also influence the timeline of the project.

1.6.3 Eco System Risks (Macro Environment)

Macro environmental risks could occur because of changes in the regulatory or overall economic situation.

Decisions of the regulatory agency have an influence on the PSTN migration process especially in the preparation phase (product development) and during the execution, for example, leading to:

- Complexity of wholesale replacement products
- Complexity of wholesale migration process
- IP products treatment (higher/lower prices, interconnection fees, etc.)
- Additional not planned regulatory requirements could happen

Even while there is a high probability that the regulatory agency will cause a significant delay in the decision-making process, the situation for each country has to be evaluated separately. All the events stated above can have an influence on:

- Prolonged migration timeframe
- Revenue loss due to different IP product treatment and interconnection fees
- Higher investment needed per closing down of each local exchange
- Possible revenue loss due to possible local exchange outages due to prolonged timeframe
- Lower cumulative savings due to prolongation of migration time and time delay of savings achievements

In addition to regulatory risks, there are other external risks as a result of the overall economic situation. For example, in price sensitive markets broadband penetration might not be reached as planned or a change in the demands of external subcontractors impacts contractor availability and prices.

Macro-environmental risks have an impact on all three major dimensions: churn, costs, and time.

In general, the financial impact should be calculated when these risks can be foreseen and have significant impact on the business case results. All risks have to be monitored permanently, reported regularly in core team meetings, and mitigate appropriately.

1.7 End-to-End (E2E) Processes

IP transformation may require changes in the processes or introduction of new ones. The main goal of this section is to list the processes that should be carefully considered and defined prior to the start of migration.

Each company has its own specifics and processes depending on the organizational structure, the level of automation and resources, and so on.

Nevertheless, based on an analysis of the processes that are implemented in the countries, some overall conclusions can be made in terms of changes in the processes due to IP transformation.

1.7.1 Adjustment in E2E Processes

In general, the processes can be clustered into three categories depending on the need for adjustment due to PSTN migration (Figure 1.5).

1.7.1.1 Processes without Major Changes PSTN migration does not have an incremental impact on the process structure and flow. But, PSTN migration should be included in these processes with high attention. Some examples of these processes include strategy and the business planning process, product management, billing and revenue assurance, customer relationship management, and CAPEX management.

1.7.1.2 Processes with Incremental Changes These processes need to be adapted, as the PSTN migration has an incremental impact on the process structure and/or flow.

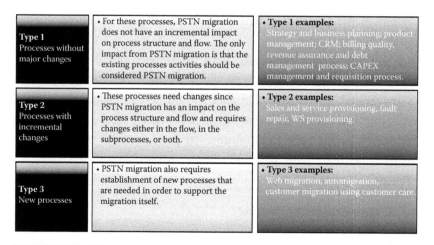

Figure 1.5 Types of E2E process adjustments.

The following processes that need changes are identified:

- Sales and service provisioning
- Fault repair
- Wholesale provisioning

For example, the service provisioning subprocess should be adapted to PSTN migration specifics such as a check of broadband access availability or installation of an iAD. The fault repair process should be adopted in a way to use this process for the migration of customers. The process changes could influence the work orders dispatching, provisioning of network prerequisites, and installation of CPE equipment.

1.7.1.3 New Processes   For the migration itself, new processes need to be defined. The introduction of new processes depends on the chosen PSTN migration type/scenario. External companies may be responsible for some parts of the customer migration and on-premises activities and, therefore, a new process is introduced for contacting the customers and on-site installation. Another example for new processes might involve automigration using the Web.

1.7.2 Complete E2E Process for PSTN Migration

In addition to considering which changes or new processes have to be adjusted for the PSTN migration, it is interesting to understand all the necessary steps that operators have to face during the PSTN migration process.

Prior to the start of the migration, it is important to fulfill several prerequisites. Drawing on the migration plan and all the variables that form it (replacement products, resources needed, lines migrated, etc.), the selected types and modes of migration are defined first. Consequently, the operator is able to define the sales potential attached to the migration and the best channel for notifying the customer about the new services.

Once all the migration prerequisites have been defined, the IMS migration—encompassing different stages—can take place. The migration varies according to the customer (RES/BUS), the stage (active/forced), and the type of migration process (ISDN/BB). However, the migration starts when a migration ticket is open in the telco's database and the customer data is sent to the appropriate business segment, which will prepare and send the information to the customer. This notification approach might be conducted through standard channels, such as leaflets, e-mails, Interactive Voice Response, and whitemailing (WG) in the case of residential customers or through customized solutions for business customers as large accounts.

Operators have an interest in leveraging less expensive migration strategies as natural migration (NM) and automigration in order to diminish the related costs, however when the customer, after several attempts, is not willing to migrate, alternative solutions such as forced migration and MSAN have to be taken into consideration. Therefore, the E2E process ends with a technical component in order to allow the IP services, which might be done by the customer himself or by a technician.

2
MIGRATION PLAN

As resources are limited, they have to be managed throughout the entire migration period by optimizing and balancing the trade-off of cost, time, and churn.

The migration plan provides this central function within the PSTN migration project while integrating all cross-functional activities and ensuring an aligned and harmonized planning process. It takes into account all possible influences which may impact the PSTN migration during the planning as well as the execution phase. During the planning phase, the migration plan ensures that all data required for an integrated approach is developed and delivered as an input into the migration plan, whereas during the execution phase the migration plan allows for quick visualization of effects from external factors, for example, resource bottlenecks (personnel as well as financial resources), thus facilitating project steering and management decisions on the force and duration of the IP migration period.

The migration plan integrates and processes input from all functional streams as described in this book. In order to avoid functional white spots, respective checklists for data delivery have been developed. It is highly recommended to use these checklists for preparing delivery of inputs for the migration plan as well as reviewing the completeness of data provided for the migration plan.

Combined with the business case framework, the migration plan provides a sound and validated basis for the operator's midterm planning phase and can be used for this purpose as well.

The migration plan is closely correlated with the business case as both tools use and process financial data. Therefore, it is mandatory that all financial data delivered as an input to the migration plan are validated and approved by the finance function prior to processing.

Description
• **Churn—avoid/minimize probability of churn**
Can enlarge due to wrong assumptions of time (e.g., prolong product development, # customer contact points) and cost (e.g., resource planning, unpredicted technical solutions)
• **Cost—manage cost**
Avoid unscheduled/unplanned investments to enable migration (e.g., MSAN POTS Card) and migration related costs (e.g., CPE, TS, CC)
• **Time—Setting/manage the correct timeframe**
Starting point and duration to finally migrate all customers by optimizing investment/savings and minimizing churn
All three criteria have an influence on realizing the potential discounted cash flow effect
Customer and area related saving potentials (e.g., energy savings per retired port, SLA/vendor cost, energy)

Figure 2.1   Trade-off cost, time, and churn.

As resources are limited, they have to be managed throughout the migration, optimizing the trade-off of cost, time, and churn (Figure 2.1).

KEY LEARNING

- Migration by area to maximize savings, and migration by product to leverage resources (cost and time).
- Start early with customized solutions and WS solutions, as they often block the clearance of areas (time).
- Maximize leverage of natural migration (NM) as it is the most resource optimal and customer friendly way for migration (cost).
- All types of resources need to be managed with trade-offs on churn, cost, and time, and check readiness to make those sacrifices (cost, time, and churn).

To ensure that learning is implemented, an approach is developed to manage resource constraints while optimizing the different dimensions. Two general types of constraints on migration need to be managed: product development resources and migration resources (Figure 2.2).

Consequently, sequencing has to be done for products and their development timing as well as for the migration itself. How each customer will be migrated (migration strategy) and when (migration sequence) also has to be defined.

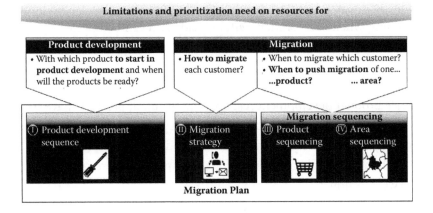

Figure 2.2 Elements of the migration plan.

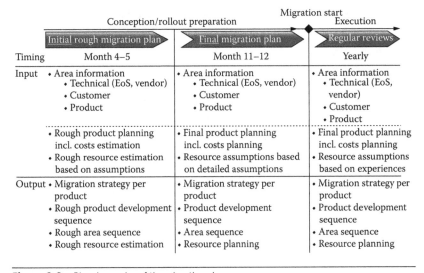

Figure 2.3 Planning cycles of the migration plan.

The migration plan is done twice in preparation for the migration start: In months 4 to 5 with rough data/assumption for an initial plan and then with finalized data in months 11 to 12. Afterward, assumptions are updated with real life experiences from migration in regular reviews (Figure 2.3).

2.1 Product Development Sequencing

The first step in the product development sequence needs to be deduced by prioritizing the products. The resulting sequence does not only determines the start point of development but also its end point, defining when each product will be ready and can be migrated.

2.1.1 Sequencing Framework for Product Development

Since product development resources are limited, products have to be prioritized for development sequence. This prioritization and sequencing of products derives from a scoring model, shown hereafter.

When prioritizing products for development, there is a trade-off between products:

- Start with wholesale and complex products as they take longer to develop and can form major roadblocks for clearing an area.
- Start with high-volume mass products to allow the maximum number of natural migrations (customer- or incident-driven migrations; cheapest form of migration).

To prioritize products for development, data on the solution complexity, effort needed, and the volume per product/segment first has to be collected from RC, BC, and WS; then, a ranking of products for development sequence can be developed (Figure 2.4).

Complexity can be driven by two dimensions:

- *Technical solution complexity*: Finding a working solution in the IP world is complex and takes a long time (especially for customized solutions and special cases).
- *Migration process complexity*: The migration process is complex and takes a long time (e.g., because it is a multilocal solution with complex immediate migration or wholesale solutions which take time to migrate due to long negotiations).

In addition to getting an early start on the complex products, operators also have an incentive to have high-volume mass products ready as soon as possible. This allows for a long period of natural migration as well as the ability to target many customers who are already in the early stages of migration. Thus, a second criterion is to consider the volume of a product, prioritizing high-volume mass products.

This criterion should be complemented by the product development effort determining the probable duration of product development.

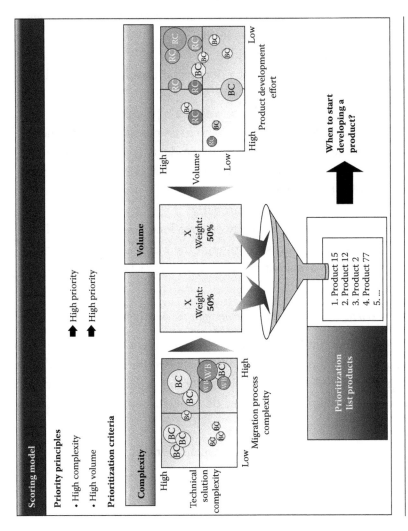

Figure 2.4 Product development sequencing.

Thus, mass products with short development types are developed first to have a substitute ready as soon as possible.

The prioritization results in different imperatives for the different product types:

1. Mass products: Push for readiness
   - They are the first to be started and also their substitutes are the first ones to be ready.
2. Niche products: Deprioritize
   - They have the lowest priority and should be deferred in favor of other products.
3. Wholesale solutions: Align with mass products
   - Due to the regulatory link to mass products, development has to be aligned with mass products and they have to be ready at an early point in time to start the long negotiation with the regulator.
4. Customized solutions: Start early
   - Need to be started early due to a long development and/or migration process. Although their development is started early, their substitutes are only ready at a later point in time.

2.1.2 Prioritization Criteria and Rationale for Niche Products

As the differentiation in prioritization of niche products gets less and less distinct, other criteria can be applied for sequencing. Where the product development criteria might not provide sufficient differentiation, additional criteria can be applied focusing on optimizing financial impact.

The financial advantage can be top-line driven:

- Higher price of the substitute than the PSTN equivalent
- Upsell potential in the new IP portfolio

Or, bottom-line oriented:

- Earlier retirement of an area
- Retirement of a PSTN platform (limited savings)

When rationalizing for niche products, two constraints should be observed:

- Product development and launch need to be aligned with the strategy and plan
- Retail and WS equivalent products/features have to be developed and introduced simultaneously

KEY LEARNING

- NRA approval before implementation.
- WS and retail service PSTN migration should be done together; possible danger for delays because WS service migration unpreparedness exists.
- Technical launch has to be prior to commercial launch. Reason: complete preparedness for migration process.
- Implementation has to be done in a nongeographical approach, as NRA can insist on no discrimination on geographical areas.

2.2 Migration Strategy (Including Phases and Types)

The second element of the migration plan (see Figure 2.2) is the definition of the migration strategy, that is, how each customer is migrated. To define the migration strategy, general timing of the migration phases need to be considered and available options of migration types need to be assessed for cost implications and customer friendliness.

2.2.1 Assess Migration Phases

Characterized by the strength of the migration push and active targeting, migration has three different phases: natural, active, and forced migration (FM) (Figure 2.5).

In natural migration, a customer has the possibility to migrate, but is currently not targeted in migration. When contact is established by the customer, through any incident (e.g., fault repair) or by non-targeted automatic migration approaches (e.g., pop-up windows), this contact is leveraged by trying to migrate the customer at the same incident. Natural migration starts as soon as product substitutes are ready and is enabled (almost) until the shutdown of the PSTN.

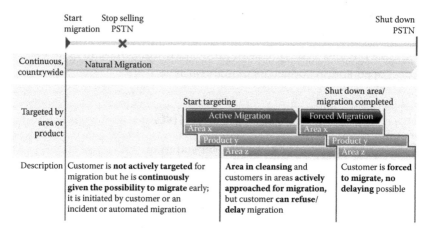

Figure 2.5   Migration phases.

As soon as the customer is actively targeted for migration, that is, contacted for migration, the active migration starts. The customer is contacted one or several times, but has the option to refuse or delay migration.

During the shutdown phase, an area has to be cleaned, thus the remaining customers have to be forced to migrate. In this phase customers cannot delay migration and can only refuse migration by terminating the contract.

While natural migration is nontargeted and countrywide, active and forced migrations depend on the target. Thus, each area or product is entering active and forced migration at different points in time.

The push for migration is increased over these phases and the applied migration strategy is strongly dependent on the phase the customer is in. The stronger the force that needs to be applied, the more costly the migration is and in addition a customer migrated in forced migration has already been targeted in active migration. Consequently, natural migration is drastically cheaper and also more customer friendly. Hence it should be leveraged as much as possible.

2.2.2 Migration Phases—Timing Guidelines and Key Learning

The migration phases differ in regard to applicability and timing. When applying the three migration phases on different types of products, it is visible that not every phase exists in full strength for each product type. While all three phases exist for standard products, natural migration cannot or is highly unlikely to be used for wholesale or customized solutions.

In the case of wholesale solutions, contact with the customer is unlikely, as he is not a direct customer but the customer of the wholesale partner. Furthermore, if given the choice, the operator would prefer to directly force the customer to migrate, as churn or customer dissatisfaction does not impact its own customer base.

Due to the necessity of technical development and testing, when migrating customized solutions, these are not subject to natural migrations. Also forced migration is likely minimized. Customized solution migrations are negotiated with business customers thus are migrated in cooperation with the customer. The operator normally gives discounts to ensure migration rather than risking loss of the customer.

Natural migration, existing only for standard products, starts as soon as the substitute product is ready. At the same time, the PSTN product should have stopped selling. All new contracts are done with the new IP product.

Active and forced migration for standardized products is done when targeting an area or a product. Their timing termination is complex and subject to the migration sequencing (see Sections 2.3 and 2.4).

The active and forced migration for wholesale solutions depends strongly on the negotiation with the regulator. As this takes up a large amount of time, it should be started well in advance. The stop selling point needs to be determined depending on the stop selling in the residential segment, as the regulator requires residential products to be also offered to wholesale partners. However, there should not be too much time between the stop selling points in both segments as the wholesale partner might take advantage of it, by selling canceled PSTN customers. This leads to additional churn but also additional migration costs, as customers are added to the PSTN by the wholesale partner, who need to be migrated later.

Migration of customized solutions requires the customer's cooperation and is characterized by a constant interaction and negotiation. It is important to contact the business customer early to gain transparency on the migration effort and complexity. When negotiating the migration plan, most times discounts are used to persuade the customer. As soon as the migration has started, no additional lines will be sold to the customer migration. The total stop of sales for a product is only reached with the start of migration of the last customer.

KEY LEARNING

- Natural migration should be fully leveraged due to the limited resources needed and should be enabled as early as possible for mass products with high usage.
- Active and forced migration should be conducted through an area-by-area approach to optimize savings and resources. Therefore, a prioritization and a sequence of the areas have to be developed.
- Customized and wholesale solutions need to be managed carefully and individually.

2.2.3 Assess Migration Types

To define how to contact the customer, the different possible ways to do so need to be assessed.

2.2.3.1 List of Migration Types Within the three migration phases there are several approaches to migrate a customer from PSTN to the IP world. Depending on the phase and the customer type, the objective is to identify the optimal migration type and scenario (Figure 2.6).

Migration Type	Description	Evaluation
Auto-migration	• Customer is requested to migrate via IVR, Web-based or call agent • Mainly for BB customer	• Cost for CC support • Minor cross-/ upselling opportunity
Supported migration	• Technician to visit customer in his premises to perform migration	• Cost for TS support • Cross- and upselling opportunity
2-in-1 migration	• Any customer or incident-driven contact is leveraged for migration attempt (esp. with technicians on-site for other reasons and migrates customer at the same time)	• No additional cost for TS support • Cross- and upselling opportunity
MSAN POTS (silent migration)	• Silent migration to POTS card, no interaction with customer required	• Cost for FS support • Cost for POTS card • No cross- and upselling opportunity

Figure 2.6   List of migration types.

Based on the experience from Hrvatski Telekom and Makedonski Telekom, there are four different types of migration used to approach the customer:

- *Automigration*: Customer is requested to self-service by migrating via IVR, Web-based, letter, or call agent. Web-based solutions can be pushed by using the "walled garden" approach, where the customer is requested to start the automigration process by recurring pop-up windows. This type is mainly valid for broadband customers, as the customer needs to have an IP-capable CPE (customer premises equipment) already. This migration type is cost efficient but does not provide the potential for cross- and upselling.
- *Supported migration*: Technician will be sent to customer premises exclusively to perform the migration. This type is mainly used for 1P customers, where broadband has to be installed and CPE to be initially set up, 2P/3P customers where automigration failed or additional migration support is required. This migration type is cost intensive but has the potential for cross- and upselling.
- *Two-in-one migration*: This type is based on any customer- or incident-driven contact, while the technician is on-site at the customer premises anyway, the customer is migrated. This type causes just minor additional costs for installation but provides cross- and upselling potential.
- *MSAN POTS*: This migration type is mainly used as a silent migration where no interaction with the customer is required. Usage of MSAN POTS cards can also be a strategic decision to migrate 1P customers instead of investing in broadband coverage. This migration type does not require additional technician support but field service support to assemble the card.

Not all of the described migration types need to be used for each phase and each solution.

One must distinguish between migration phases (natural, active, forced) and product/segments (standard products, wholesale and customized solutions).

- *Standard products*: Standard and mass products can be migrated during all phases and by using all known migration types. Exceptions are the silent migration within the forced migration phase (not affected at all using this type) and the push migration during the natural/active migration phase as this type is for forced migration only.
- *Wholesale solutions*: In principle, wholesale customers are not approached during natural migration, as they are not our own direct customers. Ideally, the operator would simply use forced migration on customers of the wholesale partner to avoid an additional cost for several migration attempts. Due to negotiations and the regulator, there might be a forced need to provide active migration as well.
- *Customized solutions*: Customized solutions for business partners are complex on one side but also main revenue drivers. To keep these customers, a time-consuming development phase for customized solutions is required. Therefore, these customers should be approached during the active migration phase only. A forced migration approach would increase the risk of churn.

2.2.4 *Define the Migration Strategy per Segment/Product and Derive the Related Costs*

In order to be able to set up a migration plan that reflects the time-related cleansing of all areas, a migration strategy has to be developed that considers the three dimensions of cost, time, and churn. This strategy defines a concept of how and when to contact the customer to ensure a successful migration of all PSTN customers.

2.2.4.1 Framework Migration Strategy for Each Segment/Product Including Unit Cost per Migration by Segment/Product A migration strategy has to be developed for the active and forced migration phase only. As a natural migration is mainly customer or incident driven, we assume that migrations within this phase are successful on the first attempt and costs can be calculated simply by multiplying the average migration cost times the number of customers.

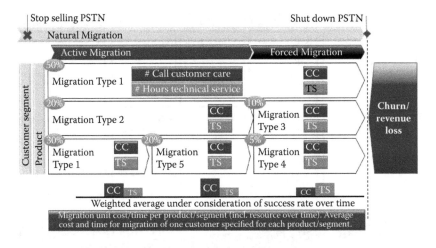

Figure 2.7 Migration strategy framework for each segment/product.

The idea of the migration strategy is to define the average cost and time for migration of one customer specified for each product/segment (unit cost) and afterward scale it to the total number of customers using the same product (see Figure 2.7).

To identify this migration unit cost for each product, as a first step all potential/applicable migration types for one product have to be considered. For each of these types, the effort in terms of number of CC calls and related cost/time, as well as number of TS visits and related time/cost have to be estimated. Considering the potential success rate of each call/visit for the different migration types, the weighted average can be calculated. In case the migration types are not sufficient to migrate 100% of the customer base, churn has to be considered.

As a base for establishing the overall migration plan, this evaluation creates transparency on how and when each customer is addressed for migration and what the related assumptions on time, cost, and churn are.

2.2.5 Migration Strategy per NatCo

Depending on local conditions, different migration strategies have been developed by NatCos extending from a pure MSAN approach to pure full IP (Figure 2.8).

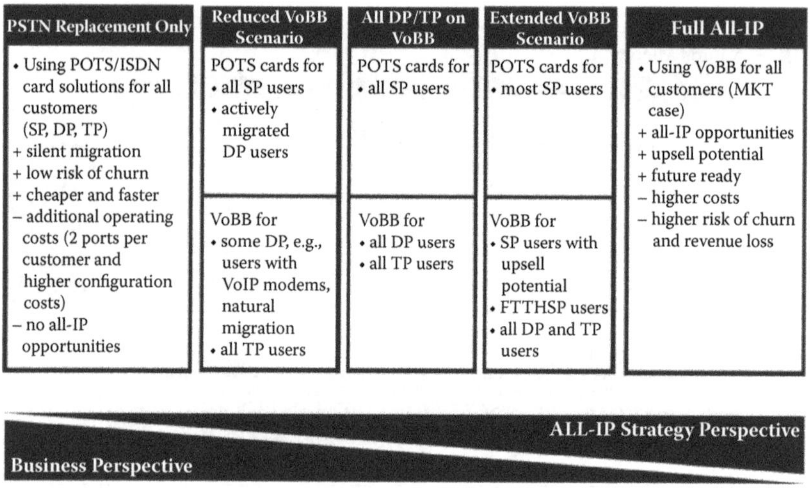

PSTN Replacement Only	Reduced VoBB Scenario	All DP/TP on VoBB	Extended VoBB Scenario	Full All-IP
• Using POTS/ISDN card solutions for all customers (SP, DP, TP) + silent migration + low risk of churn + cheaper and faster – additional operating costs (2 ports per customer and higher configuration costs) – no all-IP opportunities	POTS cards for • all SP users • actively migrated DP users VoBB for • some DP, e.g., users with VoIP modems, natural migration • all TP users	POTS cards for • all SP users VoBB for • all DP users • all TP users	POTS cards for • most SP users VoBB for • SP users with upsell potential • FTTHSP users • all DP and TP users	• Using VoBB for all customers (MKT case) + all-IP opportunities + upsell potential + future ready – higher costs – higher risk of churn and revenue loss

ALL-IP Strategy Perspective

Business Perspective

Figure 2.8 Different migration strategies in other countries.

2.3 Migration Sequencing—Product Sequencing

When determining the migration sequence, two generic approaches can be chosen: migrated product by product (PbP) or area by area (AbA).

In its extreme forms, PbP allows starting migration as soon as the first product is ready, thus resources are leveraged over a longer period of time. Also, the delay risk is lower, as a delay in product development still allows proceeding with the migration of other products.

Area by area allows areas to be cleaned earlier yielding earlier savings. However, all products for the area need to be developed and in an extreme case that can mean all products. Migration starts later, increasing the resources requirement in the shorter migration phase. Additionally, delays in product development will cause an overall migration delay.

When combining the two approaches, advantages of both can be leveraged (see Figure 2.9). Migration is started when the first product is ready and is done PbP. As soon as all substitutes for an area are ready, the migration focus is shifted to an AbA approach, cleaning the area first. Spare resources can be used further to parallel to PbP. Thus, migration resources are better leveraged with less risks of delay, while there are still savings at the earliest point in time. When prioritizing the allocation of resources over the different migration foci, the following sequence is recommended:

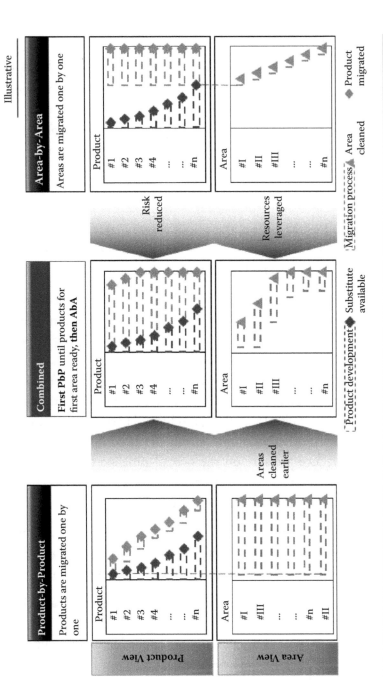

Figure 2.9 Advantages of a combined sequencing approach.

1. Natural selection
2. Area by area
3. Product by product

These results in natural migration enabled for the whole migration phase, a first phase of pure PbP followed by phases dominated by AbA complemented with PbP migration.

For both PbP and AbA, sequencing has to be done. The PbP sequencing approach is presented in the following section.

2.3.1 Prioritization Logic for Product Sequencing

In the following section, the prioritization logic for products within the product-by-product approach is provided. There are two rationales for why products should be migrated first:

- Products that show low natural migration should be migrated before products with a high likelihood for natural migration, because the one with the high likelihood is strengthened with time reducing the migration costs in comparison to the products with lower natural migration efficiency ($\rightarrow$ maximize opportunities). As natural migration does not exist with customized or wholesale solutions, it is limited to standard products.
- Products that take time to migrate need to be started early to reduce pressure on the overall timeline ($\rightarrow$ reduce delay risk). These are complex products mainly present as customized solutions for business customers.

When assessing natural migration effectiveness, assumptions on the success rates of natural migration need to be made for each product. These assumptions should take into consideration:

- Migration willingness/likelihood by customer associated with product
- Effectiveness of migration types applied in natural migration
- Ease of migration/success rate of migration

These assumptions need to be validated with experience values as soon as the migration has started (Figure 2.10).

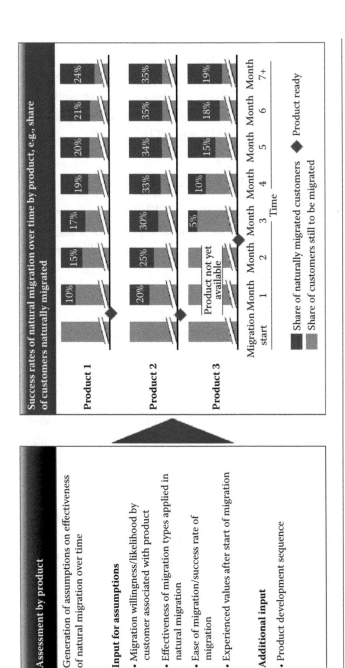

Figure 2.10 Prioritization logic for product sequencing—natural migration effectiveness.

The strength of a delay risk is defined by two dimensions: How strong is the risk (migration duration)? and How often is it inflicted (to how many areas does it apply)?

From the resulting sequence in combination with the defined migration strategy and unit costs (see Section 2.2.4), the resources needed for the pure PbP phase can be derived, as well as the customers migrated in that phase and thus the customers can still be migrated in the AbA phase[*] (Figure 2.11).

KEY LEARNING

- Standard products with a low likelihood of success for natural migration and complex products with a long duration and presence in many areas should be the first to be migrated.

2.3.2 Prioritization Logic with Criteria and Rationale for Niche Products

As the differentiation in prioritization of niche products becomes less and less distinct, additional criteria can be applied for sequencing. Rationales for prioritization should follow the resulting financial benefits. They can be divided into two groups: revenue impact and cost savings from realization of development of new IP products.

A revenue impact can be fostered in two ways: the new IP product will be sold at a higher/lower price than the PSTN equivalent or the new IP product promotes upselling.

Bottom-line impacts result from earlier savings realizations, which can be on the retirement of an area or the retirement of a PSTN platform (e.g., PSTN voice mail). The timely retirement of the targeted areas can be ensured by prioritizing the products in the targeted areas from the migration plan and for those products to start with the products that take the longest to develop.

However, the development and launch sequence has to be strongly aligned with the migration strategy.

---

[*] An even distribution over the areas per product is assumed.

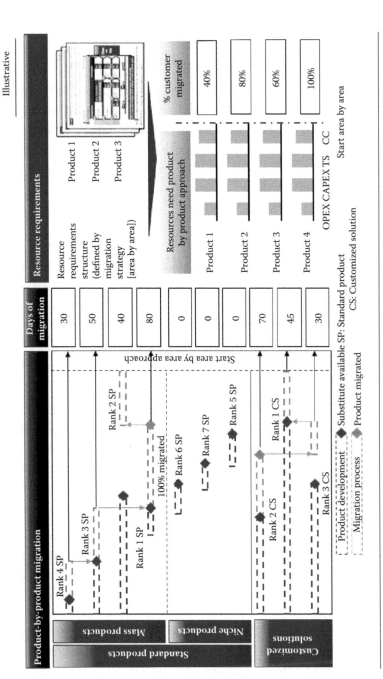

Figure 2.11 Prioritization logic for product sequencing—deduction of status at the end of pure PbP.

Another constraint that needs to be taken into account is the required simultaneous launch of retail and wholesale solutions.

2.4 Migration Sequencing—Area Sequencing

After the products are prioritized for the PbP phase, the areas also need to be planned and sequenced. In the following, an approach for area sequencing is introduced. As in the area sequencing there will be trade-offs between areas; it has to be defined which area has priority. Therefore, an evaluation approach is introduced to be applied in the sequencing framework.

2.4.1 Area Evaluation

Due to timing limitations and resource availability, the areas need to be sequenced, determining starting and end point of active and forced migration per area.

The following scoring model yields a ranking of areas used to decide in cases of trade-offs between areas, that is, it defines which areas need to be put to a timeline first, and not which to migrate first. There are two principles for high priority:

- *High limitations*: In case of strong limitations regarding timing an area is less flexible thus has priority in the timing over areas with high timing flexibility.
- *High savings opportunities*: Areas with high savings opportunities have a high priority as the financial benefit can be maximized by choosing the right location.

The criteria can be distinguished in defining the financial priority and the technological priority, which should be weighted to gain singular priority. Priority weights and weights within the criteria have to be assessed individually by each telecommunications company (Figure 2.12).

Criteria for financial priority are OPEX savings per area and required CAPEX/OPEX invest per area. OPEX savings can be area or customer driven and are consolidated to an area level. Other savings like general SLA savings are not considered. The higher the savings, the higher the opportunity, and thus the higher the priority.

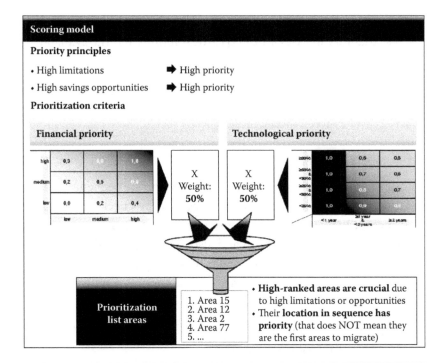

Figure 2.12 Area evaluation.

The second dimension is the incremental investment that is required for migration. As CAPEX is limited, the location and duration of the area has to be carefully planned and is not that flexible, thus the priority is higher for areas with a large investment required.

The technical priority stems from two technical limitations of timing:

- An early end of support (EoS) of an area forces the area to be migrated early. This timing constraint leads to a high priority. Instead of the end of support, the incident risk for a switch can also be used (e.g., when EoS already happened).
- Technical enablers of PSTN migration need to be present, as they determine the readiness of an area for migration. A prominent factor is the broadband coverage established by the broadband rollout plan. If insufficient broadband coverage is existent, then an area has to be placed later in the migration at a point when the rollout plan has ensured sufficient coverage. Consequently, this forms a timing restriction as to the possible starting point of the migration. The limitation

leads to a high priority. The same logic applies to MSAN port availability in case of POTS cards solutions and the availability of the general access network capacities.

The resulting ranking is used in the following area sequencing approach.

2.4.2 Area Sequencing Approach

While the areas are now ranked, they need to be sequenced in the right order and resource availability needs to be checked. To do the sequencing, it is necessary to know how many resources are needed for active and forced migration and also how long it will take accordingly. This, however, leads to circular reasoning (see Figure 2.13).

The location in the area sequence depends on the required resources and migration duration of an area,

> ...which depends on, how many customers need to be migrated in active and forced migration,
>
> ...which depends on, how many customers of an area are already migrated naturally,
>
> ...which depends on, how long the phase of natural migration for this area was,
>
> ...which depends on, where in the sequence the area was placed (the sequencing of areas).

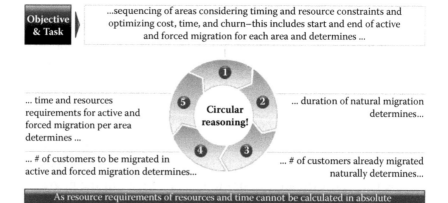

Figure 2.13   Area sequencing approach—circular reasoning.

To break this circular reasoning, a scenario approach is used. First, the assumptions on natural migration effectiveness from area evaluation (see Section 2.4.1) is combined with the number of customers per area after the pure PbP phase (see Section 2.3.1) to generate volume scenario for each area, stating how many customers are still present in each area for different migration starting points. Then for each scenario the required resources are calculated. In the last step areas are put into an initial sequence, which is then iterated for optimization and ensuring sufficient resources.

From the area evaluation (see Section 2.4.1), assumptions per product exist on how successful the natural migration will be over time. Combining this information with the number of customers per product that still exist in each area after the pure PbP phase (see Section 2.3.1) leads to a scenario depending on the starting point of active migration. It states how many customers per product are still to be migrated in active and forced migration.

The migration volume scenarios are combined in the next step with the assumption on unit cost per migration by product, which are defined in the migration strategy (see Section 2.2.4). This yields the estimated resource requirements for each scenario for different resource dimensions (CAPEX,* OPEX, regional technical service FTEs, customer care FTEs). More specifically, not only are the total resources required delivered, but also their distribution over time, assuming no resource constraints.

In the last steps the areas are sequenced. Therefore, timing constraints are derived which limit the option space for migration of each area and this eliminates some scenarios. These timing constraints can limit the starting point as well as the end point of the migration. Constraints limiting the starting point are the readiness of substitutes for each area as well as the readiness of technological enablers as sufficient broadband coverage.

While the area migration can theoretically start before all the substitutes for that area are launched, it cannot be finished. An area migration will only start if it can be finished within a reasonable time. Thus, a starting point constraint is established, even if it is slightly fuzzy.

The limitation technical enablers for PSTN migration (broadband coverage, MSAN ports, access network capacity) were already assessed

---

* CAPEX refers to customer-driven CAPEX.

in the area evaluation (see Section 2.4.1). As long as they are not given in sufficient extent again, the migration cannot be finished, leading to a fuzzy starting point constraint similar to the case of product readiness.

Restrictions on the end dates of migration are given by the end of service dates as well as by the general PSTN shutdown deadline. End of service dates can be replaced with the points of time where the risk of failure due to end of service is too high.

From the remaining scenarios a scenario per area is selected with the principals:

- Earlier scenarios for areas with high OPEX savings to maximize the financial benefit.
- Later scenarios for areas with a high cost and resource difference between early and later scenarios to reduce migration costs and effort.

Areas with high priority (see area evaluation, Section 2.4.1) should be selected first.

With the decision on the scenario, a first initial sequence is established. Furthermore, in combination with the defined migration strategy (see Section 2.2.4) and its unit costs, the required resources per area over time can be derived. This refers to the active and forced migration, but also to the required resources for natural migration as the natural migration assumptions (see Section 2.1) allow a back calculation of the number of customers migrated naturally as well as their resource needs over time.

These resource requirement profiles per area can be consolidated and complemented with further resource requirements

- as further migration-related but non-volume-driven CAPEX (e.g., platform invest)
- as technical service and customer care FTE requirements for regular task or migration-related tasks (e.g., migration-driven fault clearance)

This leads to a total resource profile over time for each resource dimension.

In the final step, the resources profiles of the initial sequence are mapped to available resources to identify mismatches. In the following, the sequence needs to be iterated to ensure sufficient resources at each point in time. In addition, there are further optimization options which include influencing the timing constraints, the available resources, and the resource requirement. However, each option has effects on time, cost, and churn (Figure 2.14).

The basis for resource planning is the following information:

- Targets of the PSTN migration project/resources/deadlines.
- Detailed analyses of the regular processes versus number of current FTEs involved in the affected processes and areas (e.g., field operations, network and service operations, customer care, marketing).

Then the detailed resource planning should be based on the following information:

- Plan for realization: Migration through fault clearance and provisioning, IMS migration flows—services with HGW, IMS migration flow—PSTN and PSTN+ADSL (modem), network and service operation (N&SOD) additional resources, migration through the call center.
- Additional activities arising from PSTN (type, quantity, duration, complexity, etc.).
- Number of FTEs with appropriate competencies, which can be used in the PSTN migration process (in parallel with their everyday activities).
- Gap (activities and FTEs), which should be covered with external human resources.

The spare resources identified in the final sequence can be leveraged for PbP migrations.

The migration sequencing model contains some simplifications in the form of assumptions. However, the risk is limited, resulting in a perhaps not optimal sequence but in the second best solution ensuring sufficient resources (Figure 2.15).

KEY LEARNING

- Due to the interdependency of sequence and resource requirements, a scenario approach is necessary.
- Crucial areas need not be targeted first but set as the best location in the sequence.

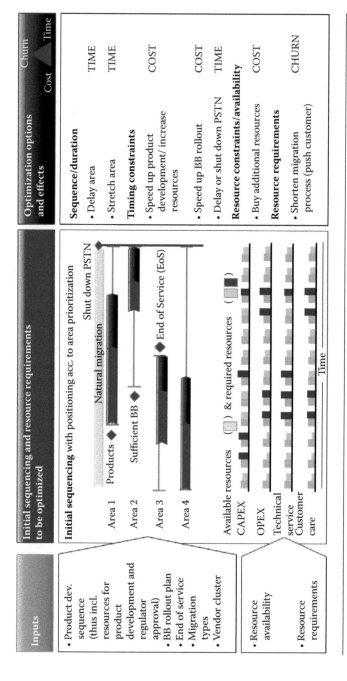

Figure 2.14 Area sequencing approach—optimization of initial sequence.

Simplification	Assumption	Risk	Risk Assessment
Share of migrated customer by PbP is applied evenly on all areas	PbP migration impacts all areas at the same rate, i.e., is evenly distributed	The resource requirement of areas is changed and thus the sequence	In case of small # of migrations limited effect, high # of products of large number ensures even distribution → LOW
Impact of natural migration on volume for product migration is not considered	Impact of natural migration on PbP is negligible	Too high resources and time requirements planned	As product sequencing focuses on standard products with low natural migration, effect is limited → LOW
Resources and impacts of product migrations in mixed phase are not considered	Resource/time savings due to reduced volume in AbA migration due to PbP migrations are leveling out with resource requirements for PbP migrations	Estimated resources requirement is in beginning of AbA too low and later too high Optimization on real values might lead to different results Timewise it levels out	Result will be only 2nd best solution but still optimized In addition, it is a more conservative approach and includes a natural risk buffer → MEDIUM

More conservative model as a 2nd best solution of optimization

Figure 2.15 Area sequencing approach.

3

PRODUCT PORTFOLIO ROADMAP

A product portfolio roadmap describes the necessary process and steps for transferring the PSTN products to an IP world.

The approach for the IP product portfolio definition is structured into four steps:

1. *Information collection*: Gathering of information of features (availability, necessity) and regulatory requirements.
2. *Analysis and selection of features*: Selecting the features to be realized within the IMS world based on a decision framework and analysis by feature.
3. *Product definition*: Defining new products from the selected features including term and condition changes and NRA announcement.
4. *Realization of a new portfolio*: Developing and launching of a new product.

3.1 Information Collection

This step collects information that is required in the next step, about which features should be realized in the all-IP product world. Within this step, there are four activities that need to be done:

1. Collecting a list of complete current features/product catalogue/terminal equipment:

   The first activity is to gain transparency on existing products, deriving the PSTN feature list and collecting the information on the terminal equipment used on the customer side.

2. Preparing a regulatory checklist:

   A list of regulatory-required features is derived from all legal and regulatory obligations.

3. Identification of technical available features:

NT/IT deliver the available features on the IMS platform within three categories ("already ordered," "off-the-shelf available," and "can be developed").

4. Matching the current features to the available IP features:

Information for PSTN features, mandatory features, and availability of features are combined to identify the features that will be transferred 1:1, with an investment of those to be assessed and those to be retired.

The output of the information collection is a categorization of features into:

- Mandatory features
- Features in discussion for investment (assessment in step 2)
- Features that cannot be replicated
- New IP features

3.1.1 A Collection List of the Complete Current Features/Product Catalogue/Terminal Equipment

To start to transfer the PSTN products to the IP world, you first need transparency on the existing PSTN products, features, and terminal equipment used.

Starting with the PSTN/ISDN product portfolio before the start of PSTN migration, the list of PSTN features is derived.

The technology migration requires full attention, not only on the main product characteristics, but also on the whole product portfolio built upon it, as well as "hidden" usages such as faxes, POS terminals, ATMs, and so forth. There are features and terminal equipment that can cause severe problems and thus their solutions are crucial.

In order to cover all these issues, the following items are taken into account:

- The PSTN/ISDN product portfolio before the start of the PSTN migration

- Mandatory regulatory services (number portability, CS, CPS, etc.)
- Hidden usage of the telephone lines

During negotiations with the vendors, and later, during the implementation and testing, several problems with services and features, so-called special cases, are identified. These special cases can be divided into two groups according to the problem; on the one side is the terminal equipment; and on the other, the central office equipment.

3.1.1.1 IP Migration of ISDN—ISDN Features Missing in IMS It is not possible to replicate all ISDN features in the new all-IP world. Therefore, it is important to perform an assessment of which services and features are used by customers. After the investigation is completed, an evaluation can be performed to determine which of the services and features require an alternative solution and which ones will be terminated.

KEY LEARNING

- Make sure to have a complete list of products and features.
- Be prepared to identify the issues; involve all relevant areas (sales, marketing, regulatory, NT/IT, WS) and vendors.
- Have a process/approach on how to manage them.

3.1.2 Preparation of a Regulatory Checklist

Some features of legal and regulatory obligations are required to be provided in the new IP world as well. The features need to be collected and their implementation ensured.

In this phase, the required features are gathered from

- Electronic Communications Act
- All appending bylaws, reference offers
- Universal service obligations (USOs) (prescribed by the act and bylaws on USO)

The main target is to determine the list of regulatory-required features.

Table 3.1   List of Regulatory-Required Features

REGULATORY-REQUIRED FEATURES	COUNTRY 1	COUNTRY 2
Detailed call log	Yes	Yes
Outgoing call restriction—fixed (OCBF)	Yes	Yes
Outgoing call restriction—variable (OCBV)	Yes	
Calling line identification restriction—onetime (CIDBLK)	Yes	
Calling line identification restriction—permanent (CLIR)	Yes	Yes
Secret number—dialing directory	Yes	
Billing amount limit (BL)	Yes	
Individual outgoing call barring (IOCB)	Yes	
Call diversification protection (CDP)	Yes	
Number portability (NP)	Yes	Yes
Carrier selection/Carrier pre-selection (CS/CPS)	Yes	Yes
Anonymous call rejection (ACR)	Yes	Yes
Connected line identification presentation (COLP)	Yes	Yes
Connected line presentation restriction (COLR)	Yes	Yes
Malicious call identification (MCID)	Yes	Yes
Temporary barring outgoing call (TBOC)	Yes	
Calling line identification presentation (CLIP)	Yes	Yes
No CLIP	Yes	
CLIP restriction override (CLIPRO)	Yes	Yes
Incoming blocks—reject all incoming calls		Yes
Lawful interception	Yes	Yes
Emergency calls (ECs) 24/7	Yes	Yes

In addition, problematic issues related to the product portfolio should be detected. Argumentation for the problematic issues should be prepared (in case the regulator opens a discussion on the problematic issues).

In Table 3.1, the list of the regulatory-required features in two countries as well as a list of additional regulatory requirements impacting the product development and PSTN migration are provided.

3.1.2.1 List of Regulatory-Required Features   The features in Table 3.1 were required by the regulators in two countries.

3.1.2.1.1 Argumentation for the Problematic Issues   When preparing for PSTN migration, certain problematic issues related to service features might arise. For example, IP-based voice service will be unavailable in case of a power outage at the end-customer's location; also, the provision of IP-based voice service imposes new network elements, which might interfere with the existing cost structure and thus have an impact on prices.

KEY LEARNING

- The operator should evaluate the national legal framework in order to decide how to approach the regulator. In some cases, it may require a formal notification about all the technology changes and following prescribed procedures. In any case, it is recommended to briefly inform the regulator about the process prior to the start of the PSTN migration.
- Similarly, customers should be adequately informed about the migration and impacts that migration has on service.
- Change of service features should be described in T&C/service specifications. These changes usually require a notification to the regulator (prior notification or approval) and the obligation to inform the customers about the changes in T&C.
- If regulatory requirements change during the process, they should be designed to match the provisions for IP-based services.
- Promotional campaigns to stimulate customers to migrate to all IP might be judged negatively by the regulator for the potential spillover effect on the market power and as potentially discriminating (limited to certain geographical areas).

3.1.2.2 List of Further Regulatory Requirements and Their Impact   When preparing for a PSTN migration, as well as for more general IP migration (with the aim to reduce costs), additional regulatory requirements/issues might arise (Table 3.2).

3.1.3 Identification of Technically Available Features

To assess which of the existing PSTN features can be realized in the IP world, NT/IT need to provide a list of available features, which are:

- Existent and already ordered
  - Due to TISPAN
  - Due Deutsche Telekom frame contract

- Able to be obtained with some investment
  - Bought off-the-shelf from supplier
  - Able to be developed

3.1.4 Matching of Current Features to Available IP Features

In order to identify the feasibility of a transfer of a PSTN feature to the IP world, it is necessary to map the existing PSTN features to the

Table 3.2   List of Further Regulatory Requirements and Their Impact

ACTION	REGULATORY IMPLICATIONS	TIMING
ISDN replacement	Retail product development (SIP trunk) and notification to NRA Substitute for WS services on ISDN (WLR)	Prior IMS migration
Reduction of local IC points	Announcement of reduction plans Reference offers amendments (RIO and WLR). IC price negotiations with regulator	6 months prior to reduction of first local IC point
Choose migration strategy; two possible scenarios:		Prior IMS migration
1. Use of both MSAN and BB line	WLR provided only on MSANs (not available on BB lines)	
2. Use only BB lines for IMS	WLR provided over BB lines, CPE delivered to end customer with operations and maintenance The same CPE will be used for BSA services sharing model	

KEY LEARNING

- Proactively manage additional regulatory issues with the regulator (WS/retail).
- Initiate an internal project for ISDN replacement prior to/ in parallel with PSTN migration. In any case, the ISDN replacement product must be available prior to dismantling the first IC local point.
- Initiate negotiations with regulator regarding local IC points reduction/IC price that would apply after migration to regional IC, as soon as the reduction plan is determined.

available features in the IP world. The realization of the feature in the IP world can be on the IMS platform, other platforms (e.g., from mobile), or within the CPE. This mapping has to be done thoroughly, as small differences in functionality can exist.

For the matching, several inputs are required: a list of existing PSTN features and technical feasibility of the features. Also helpful as supplementary information are regulatory-required features and availability of IP features, that is, is it already ordered, has to be bought, or developed.

The following section describes a framework for mapping the features and deriving conclusions based on implications and further next steps.

3.1.4.1 Framework for Matching Features and Drawing Conclusions about Features As a first step, the list of PSTN features (see Section 3.1.1) is mapped to the list of available features in the IP world (see Section 3.1.3), that is, already ordered or can be bought off-the-shelf (Figure 3.1).

This can result in three cases:

a. Features available in the IP world: Features that can be mapped.
b. Features not (yet) available: Not feasible or have to be developed.
c. New IP feature: Did not exist in PSTN world.

This categorization can be enhanced with supplementary information of regulatory-required features and availability/feasibility information

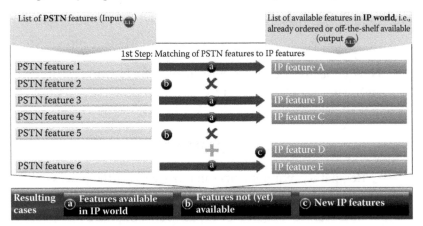

Figure 3.1   Matching PSTN features to IP features.

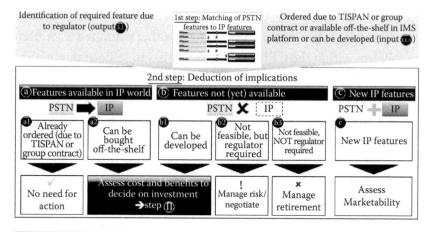

Figure 3.2 Deduction of conclusions from feature mapping.

of IP features (i.e., already ordered, can be developed, not feasible) (Figure 3.2).

From the resulting cases, action can be deduced:

a. Features available in the IP world:

1. Already ordered: No need for action
2. Can be bought off-the-shelf: Investment needs to be assessed

b. Features not (yet) available:

1. Can be developed: Investment needs to be assessed
2. Not feasible, but regulatory-required: Manage the risk and negotiate with the regulator
3. Not feasible, not regulatory-required: Manage retirement

c. New IP feature: Did not exist in the PSTN world: Marketability needs to be assessed

KEY LEARNING

- Regulatory requirements can change along the process.
- A Request of Quotation (RFQ) needs to not only consider features but also anticipate traffic implications.

3.2 Analysis and Selection of Features

For the features under discussion (output in Section 3.1.4), a decision has to be made to invest or retire. In order to select the features for the IP product, the portfolio should be based on a comparison of benefits of the feature and the costs of its realization. This analysis is done and realized through four activities:

1. *Identification of the feature value*

   The benefit or value of the feature is derived from different criteria and condensed into one value.

2. *Calculation of the costs for the features to be potentially developed/ bought*

   The cost and/or resource requirements per feature are assessed.

3. *Evaluation of features*

   Cost and benefits are combined into one framework resulting in a prioritization of features and a decision on what features to develop.

4. *Retirement of features*

   The features not developed or that are generally not feasible have to be retired.

The outcome of this step is the list of features to be implemented in the IP world, the required costs, and the retirement of features that are not realized.

3.2.1 Identification of Feature Benefits

In order to select the features to be realized in the IP world, an assessment of benefits to cost/required resources has to take place. Because not every feature generates separate revenue, other factors such as the importance to the customer have to be considered and condensed into one value for the feature benefit.

   For gathering information on feature benefits, the top-line impact (e.g., direct revenues), usage, and marketing importance per feature

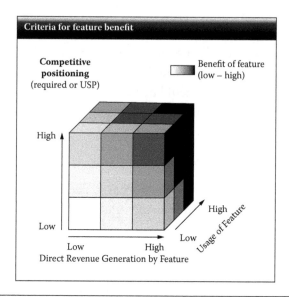

Figure 3.3 Criteria for feature benefits.

need to be available. The relative importance of different criteria has to be decided as well.

This activity provides a value for each feature, which can be combined with the costs per feature to be used for prioritization and feature selection.

To support this activity, an evaluation framework for feature value including possible criteria and approaches is presented.

3.2.1.1 Evaluation Approach for the Feature Benefits A feature can be important due to a multitude of factors which are not necessarily complementary (Figure 3.3).

The major criteria are:

- *Direct revenue*: The feature creates direct revenue, that is, it is sold separately as a value-added service and is distinctly charged.
- *Usage*: The feature has high usage by the customer indicating high importance to the customer.
- *Competitive positioning*: It is important for selling the products as it is either viewed as a basic feature and expected to be available by the customer or it is a unique selling point and creates a differentiation to other operators.

The importance of a feature results from a combination of these factors.

It can be shown graphically in a coordinate system. The relative importance of each criterion, and its weight, can be presented by the length of each axis. By defining the relative axe lengths and the position of each feature in the system, the feature value is indicated by its distance to the origin.

The operational solution is a weighted calculation where for each feature the criteria values have to be defined and consolidated through predefined weights.

3.2.2 Calculation of Costs/Resources for Features in Discussion

After defining the benefit of each feature, the cost side of each feature needs to be assessed.

These costs have to include all relevant capital and operating costs. Specifically for the features, the following costs have to be checked:

- Vendor's price for feature delivery
- NT and IT costs for putting the feature into operation
- Supporting vendor's costs
- Licensing costs
- NT and IT operating costs

This activity has to result in a list of costs per feature. In the case where the feature is customer dependent the cost has to be calculated on a per customer basis.

For the regulatory-required features apart from the calculated costs, the penalty risk cost should be considered.

KEY LEARNING

- The benefit of the feature has to be carefully assessed.
- Not every important feature necessarily generates direct revenues.
- Be careful to look at usage and not at the activation status.
- Analysis can also be used as a starting point for marketing activities and promotions.

3.2.3 Evaluation of Features

The previously assessed benefits (see Section 3.2.1) and costs (see Section 3.2.2) per feature need to be set in relation to one another, to gain a prioritization of features and to select the features which are beneficial to implement.

3.2.3.1 Evaluation Framework for the Selection of Features The output of the previous activities, benefits (output in Section 3.2.1), and costs (output in Section 3.2.2) per feature can be entered in a matrix. In addition, the differentiation of regulatory-required features and optional features (requiring an investment assessment) should still be considered (Figure 3.4).

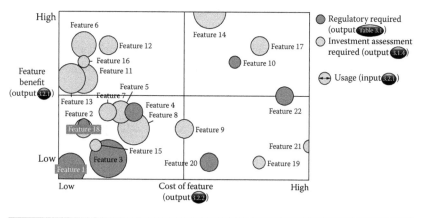

Figure 3.4 Benefit-cost matrix for the selection of features.

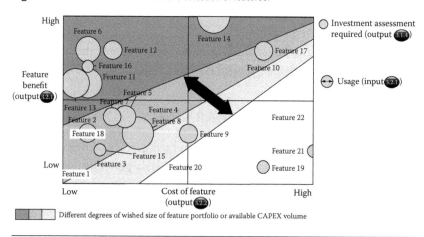

Figure 3.5 Evaluation framework for the selection of features.

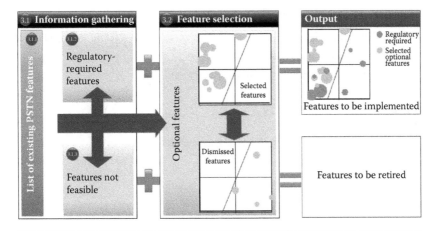

Figure 3.6 Identification of features to implement and features to be retired.

In the next step, optional features are differentiated into features to implement (positive benefit-cost relationship) and into dismissed features (negative benefit-cost relationship). The location of the threshold depends on the preferred size of the portfolio as well as on the available CAPEX for implementation (Figure 3.5).

Combining the information from step 1, regulatory-required features (see Section 3.1.2) and not feasible features (see Section 3.1.3) and results from the feature selection on optional features (see above) yields the list of features to be implemented as well as the list of features to be retired. Furthermore, the costs of implementation can be derived (Figure 3.6).

3.2.4 Retirement of Features

Features that will not be realized in the IP world need to be retired, requiring actions and communication in accordance to regulatory obligations.

The features to be retired are defined by the identification of the features that cannot be replicated in the IP world and the decision of which features will not be implemented. Regulators require a specific process for the retirement, which needs to be requested.

After communication with the regulator about the retirement and the compliance of the operator and regulator-specific processes, the features in question are retired (Table 3.3).

Table 3.3   Example List of Terminated Features

RETIRED (TERMINATED) FEATURES	COUNTRY 1	COUNTRY 2
Additional number with different ringing		Yes
Awakening (alarm call), automatically and semi-automatically		Yes
Counter at customer		Yes
Hot dialing	Yes	Yes
Hot dialing, prolonged		Yes
Sub-addressing—telephone line 10, 20, and 30 voice circuits		Yes
Terminal portability (also for WS)		Yes
Absent customer		Yes
MSN numbers		Yes
Home SMS	Yes	
Do not disturb	Yes	
Speed dial	Yes	
Last number redial	Yes	

3.3 Product Definition

Based on the defined portfolio of features within the IP world (output in Section 3.2), the new product portfolio needs to be defined and necessary adjustments, for example, to the service specifications, need to be realized. Therefore, three activities take place:

1. Feature mapping to the existing portfolio:

   The existing PSTN product portfolio is mapped to the IP features that will be realized.

2. Substitute the product definitions:

   The mapped portfolio is adapted to define the new products and its details.

3. Announcement to the NRA and changing of the T&C and the submission of the RO:

   The new products might require changes in different documents (e.g., terms and conditions, service specifications, pricing schemes), which need to be approved by the regulator.

3.3.1 Feature Mapping to the Existing Portfolio

As a starting point for defining the new product portfolio, decided IP features (output in Section 3.2) are mapped to the old PSTN products,

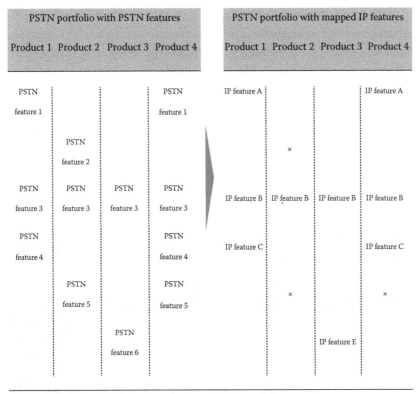

Figure 3.7 Example of feature mapping to an existing portfolio.

providing information on how the PSTN would look in the new IP world if no further adjustments were to be done.

For this activity, the features of the PSTN product portfolio are checked for replacements or if they are retired, leveraging the matching of PSTN features to IP features (see Section 3.1.4) (Figure 3.7).

3.3.2 Substitute Product Definition

After the selection of IP features, these need to be structured into the new IP products. The old PSTN product portfolio mapped to the selected IP features (see Section 3.3.1) serves here as a starting point.

Afterward, the product definition should not only include the product itself but also the proposition, its pricing, the cost modeling for regulator negotiation, and so forth.

Due to potentially different features, changes to the product portfolio might be necessary or beneficial, for example, to increase transparency for the customer experience or simplification/modularization.

To define products ensuring a transparent, simple, value-optimized, and differentiated portfolio, different factors can be taken into account, which are presented in the following section. It further supplies an approach on an NPV assessment.

3.3.2.1 Factors Impacting Decisions on the Product Portfolio Definition There are different factors to take into account when deciding if and how to adapt the product portfolio (Figure 3.8).

Customer experience and simplicity comprise internal benefits as well as customer experience-based advantages (Figure 3.9).

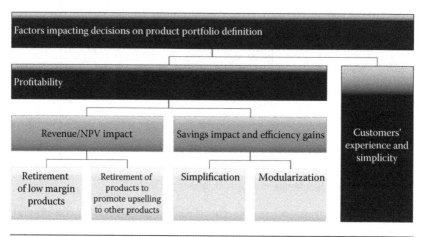

Figure 3.8 Factors impacting decisions on a product portfolio definition.

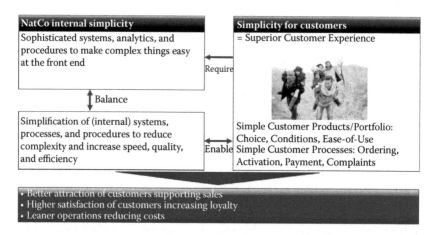

Figure 3.9 Advantages of product portfolio simplicity.

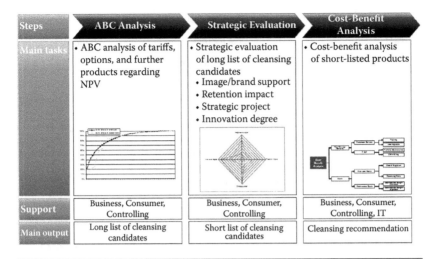

Figure 3.10   Approach for product cleansing.

3.3.2.1.1 Approach for NPV Assessment   When assessing the profitability of the product portfolio, a three-step approach is recommended (Figure 3.10).

KEY LEARNING

- Proactively shift customers to products that reduce complexity, enable margin improvements, and cost savings.
- Phase out products that have a negative contribution margin and low usage.
- Limit development of products that are not aligned with the product portfolio.

3.3.3 Announcement to the NRA and Changing of the T&C and Submission of the RO

Due to PSTN migration, T&C, service specifications, and reference offers might have to be amended to reflect technological and feature changes. Depending on the national regulatory framework, amendments of T&C, service specifications, and reference offers have to be notified to the regulator/approved by the regulator. To support this activity, a checklist is provided with potentially required regulator approvals and some best practices in T&C.

Table 3.4   Checklist for Potential Regulatory Approval and Best Practices in T&C

ACTIVITY	DESCRIPTION	TIMING
Changes of T&C	Description of IP technology features	
	Impacts on quality of voice services (power outage)	
	Enable network modernization (customer shall be obliged to follow instructions necessary for network upgrade)	Prior to IMS migration
	Include the right to terminate the subscriber agreements (USO agreement if the customer fails to enable modernization)	
Changes of services specifications (subscriber standard agreements)	Point out that voice will be migrated to IMS	Prior to IMS migration
	Technical description of the CPE (to match IMS requirements)	
Changes of reference offers	Technical rep location of MS features on WS offers (amended WLR or BSA offers)	Prior to IMS migration/when WS obligations are imposed
	If WLR provided over BB lines, provision of CFE for the WS customers	

3.3.3.1 Checklist for Potential Regulator Approval and Best Practices on the T&C   A list of a required communication with/approval from the regulator is provided in Table 3.4. Depending on local regulatory obligations and the changes that need to be realized, approval from the regulator may become necessary.

KEY LEARNING

- T&C have to include a paragraph allowing network modernization (additional obligations on the customers' side) and to define products as technology independent.

3.4 Realization of the New Portfolio

The newly defined product portfolio has to be realized and introduced.

1. Prioritization of the development and launch sequence:

   As the realization of the new portfolio requires lots of resources, the products have to be prioritized into a development and launch sequence.

2. Planning of the launch and execution:

   After the product is developed, it has to be launched.

3.4.1 Prioritization of the Development and Launch Sequence

The prioritization of product development takes place within the generation of the migration plan in the work stream of a commercial roadmap.

3.4.2 Planning of the Launch and Execution

After defining the products and setting of the launch sequence, actual launch planning and launch have to be realized following the operator-specific go-to-market processes. After the launch planning and realization phase, all IP products should be introduced.

The subsequent migration of the customer to the new products is part of the migration plan (see Chapter 2).

3.5 Commercial Opportunities in B2B and B2C during and after Migration

The PSTN migration, as a crucial part of an all-IP migration, provides a new platform for commercial opportunities. These opportunities can be brand new services and functions as well as enhancements to existing services and functions.

A list of these opportunities in B2B and B2C is provided in Figure 3.11.

Value for customer	New IP services/products	Improved processes

• Instant provisioning and faster customer service
• Network stability
• Clear voice quality
→ Product examples:
   • Hybrid access
   • HDVoice
   • Broadband on demand (e.g., Macedonia)
   • On-demand scalability (e.g., number of lines)
	• Accelerated time-to-market
	• Future-proof product concept
	• Enables convergent solutions
	→ Product examples:
	   • IP Centrex/Hosted PBX
	   • Unified Communication
	   • IP Phone Portfolio
		• Enables Zero Touch (provisioning; repair)
		• Increase self-provisioning capabilities

Figure 3.11   Commercial opportunities in B2B and B2C.

4
NT/IT Roadmap

The NT/IT roadmap consists of three substreams:

1. Target architecture and technical product development:

   Technically developing new products and deriving the NT target architecture.

2. NT rollout:

   Establishing NT readiness for rollout, planning resources, and preparation for rollout.

3. IT roadmap:

   Establishing the readiness of IT applications, especially implementing new products and processes (i.e., implementing a migration process and adjusting existing processes).

In addition, the NT/IT roadmap might have high interdependencies with other NT/IT projects (e.g., Next-Generation Customer Relationship Management [NG CRM]), whose management is the subject of the last work substream (Figure 4.1).

4.1 Target Architecture and Technical Product Development

This substream covers the definition of the technical target architecture and subsequent technical product development implementing the new products.

For the technical target architecture, the respective strategy documents are collected and the target picture is defined.

I. *Network architecture*: Preparation and evaluation of the overall network in order to define the network development steps and to prepare the future network architecture.

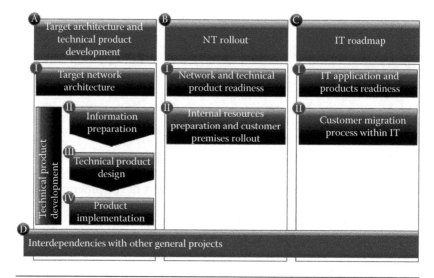

Figure 4.1   Substreams of an NT/IT work stream.

The technical product development consists of three steps:

II. *Information preparation*: Gathering of information about the technical available features, differences in the service ergonomics, and required data for preparation of End-to-End technical solutions for basic voice service.

III. *Technical product design*: Identification and definition of all the necessary steps, inputs, and outputs connected with the technical product design process.

IV. *Product implementation*: Identification and definition of all the necessary steps connected with the technical product implementation process.

4.1.1 Target Architecture

The collection and comparison of network strategy documents consist of:

- Preparation/elaboration of the overall architecture strategy
- Determination of the network architecture development steps
- Network architecture design options

The target architecture needs to be developed for the new IP infrastructure.

4.1.1.1 The Collection and Comparison of Network Strategy Documents As the network architecture is strongly impacted, the overall structure of the future network needs to be defined. The next section delivers an approach for how to do this.

4.1.1.1.1 Approach to Reach the Overall Structure of the Future Network The following passage provides an approach for developing the target picture (Figure 4.2).

Overall strategy: The preparation of voice strategy (PSTN migration) is the first step in the process of the IP transformation. It should show for the main path options how and when to reach the final goal; the "all-IP" network.

The following preconditions should be considered:

- A detailed elaboration of the status of the existing PSTN network and provided services.

  PSTN migration cannot be isolated from the broadband transformation, since users need to have access to all services (including voice) over one common broadband access. Therefore, broadband penetration is one of the main prerequisites for voice migration and could determine its timeline.

- End of support of the existing legacy switching systems.

  This is crucial because it is the main reason for PSTN migration and the risk of failure increases day by day.

- Technical preconditions and a timeline for their realization.

  The overall structure of the network has to be clearly defined, elaborated, and harmonized before starting the migration.

Architecture development. The architecture development steps concern the following topics:

- Access level transformation roadmap (single port per customer)
- Transport level integration roadmap
- Development concept for unified core control and application level

Steps	Overall strategy	Architecture development steps	Design options
Description	• Preparation of voice strategy (PSTN migration) is the first step in the process of IP transformation. It should give main directions on how and when to reach the final goal, "all-IP" network.	• Access level transformation roadmap (single port per customer) • Transport level integration roadmap • Development concept for unified core control and application level	• Description of the key points of differences • MSANs and only BB • BCS, AS on Core IMS • Chaining and/or services orchestration • IN services – SDP or existing platform
Prerequisites	• Detailed elaboration of existing PSTN network status and provided services • End of support of the existing legacy switching systems • Technical preconditions and timeline of their realization	• Broadband penetration • Which types of CPE devices to be used in the process of PSTN migration • Dimensioning to be done in a way to satisfy traffic demands depending on bandwidth requirements per service as well as number of users per services • IP transport: VLANs configuration	
Dependencies	• Current voice product portfolio vs. future voice product portfolio		
Result/Milestone	• Overall structure of future network	• Target network architecture picture	• Possible different approach to the same issues/solutions • Described on Annex

Figure 4.2 Approach to reach the overall structure of the future network.

The following preconditions should be considered:

- Broadband penetration
- Types of CPE devices to be used in the process of PSTN migration
- Dimensioning to be done in a way to satisfy traffic demands depending on bandwidth requirements per service as well as number of users per service
- IP transport: VLANs configuration

Design options. In the final step, the different design options are identified and evaluated.

From a technical perspective, IP transformation is the migration toward standard-based architectures that allow service providers to create multipurpose platforms sharing a common infrastructure.

A general overview of the evolution of legacy PSTN networks toward next-generation networks (NGNs) and IP multimedia subsystems (IMS), as well as the target network, is provided in Figure 4.3.

KEY LEARNING

- The overall structure of the network has to be clearly defined, elaborated, and harmonized before starting the migration.
- The target network architecture has to be determined and all the prerequisites need to be fulfilled before the decision for future development steps.

4.1.2 *Technical Product Development—Information Preparation*

This task is responsible for collecting information about the available technical features, the differences in the service ergonomics, and the required data for the preparation of End-to-End technical solutions for the basic voice service. This step consists of three activities:

1. End-to-End (E2E) technical solution for basic voice service.

   The first activity is to prepare the simplified End-to-End solution schemes of standard PSTN/ISDN2/ISDN10/20/30

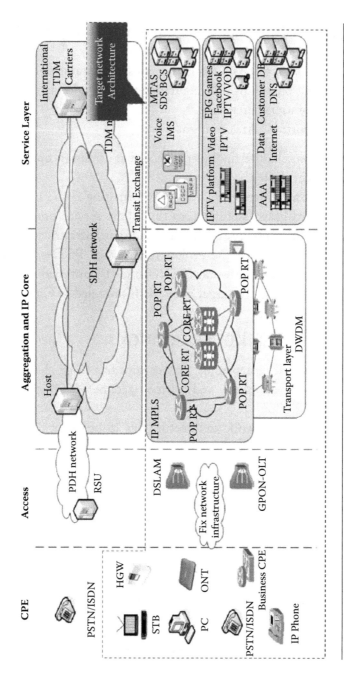

Figure 4.3 General overview of the evolution of the legacy PSTN network toward NGN and IMS.

migration scenarios for basic voice service covered by the MMTel (Multi-Media Telefone) application server.

2. Preparation of the list of available IP-based technical features.

   Mapping a list of PSTN features versus available IP-based features in order to provide an overview and awareness of the technical availability to the marketing and regulatory departments. Available technical features on the IMS platform are delivered within three categories ("already ordered," "off-the-shelf available," and "can be developed").

3. Identification of the differences in service ergonomics due to the introduction of new IP technology.

   Information for improvements of the service ergonomics due to an enhanced feature solution implemented on the IMS platform and for the complexity in service ergonomics due to CPE limitations.

4.1.2.1 End-to-End Technical Solutions for Basic Voice Service When starting the information preparation about the technical availability of IP-based services, it is recommended to prepare simplified schemes for standard migration scenarios in order to present the End-to-End solutions of basic voice service in a simplified, understandable way.

The technology migration requires full attention not only to the main product characteristics, but to the whole service built upon it, the type of the access technology as well as the terminal equipment on the customer side, and so on.

4.1.2.1.1 End-to-End Technical Solution Pictures Clear and understandable technical E2E solution schemes for the standard migration process of all basic voice services are also required for discussions and explanations, including sections without specific technical background, such as for sales and regulatory departments.

For completeness, the following aspects should be explained:

- Service which should be assigned to the IMS user on MMTel
- Type of access technology
- CPE devices needed, tested, and available

During the process of defining the End-to-End solution for voice service, the following technical preconditions regarding the voice service should be considered:

- Customer access line requirements (capability, minimum upstream, and downstream bitrate).

  Voice services: Minimum DS/US bitrate; for two channels: 256 kbps/256 kbps.

- Transport network capabilities and QoS classes, which will enable traffic prioritization to support transport of the signals with the required quality.
- Voice channels per access line: While PSTN analogue telephone lines can provide one voice channel only, ISDN can provide from 2 up to 30 channels per line, depending on the ISDN type.

  In the IP network, in order to transmit voice, at first the signal is digitized, assembled in IP packets, and afterward sent into the network. The number of packets that are generated per second depends on the used codec for digitalization/compression of the voice. Therefore, the number of simultaneous voice channels that can be carried over one IP line (for example, ADSL) is limited by two factors:

  - Voice codec and
  - Access line throughput.

  The required bitrate for a voice channel depends on the selected codec. Figure 4.4 shows some typical codecs.

The value of the required bitrate shown in the figure is only for one direction and with a 20-ms packet assembly delay. For shorter packet assembly delay, the required bitrate increases.

In case of an ADSL, due to its asymmetric nature, the limiting factor for the number of simultaneous voice channels is the upstream speed rate. For example, an ADSL line with a 384 Kbps upstream bitrate (as shown in the graph in Figure 4.4) for a G.711 codec is able to transmit three simultaneous voice channels (theoretically

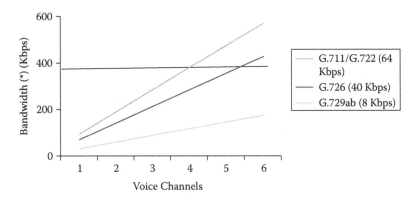

Figure 4.4   Required bandwidth per different codec.

four, but there is also some demand from other protocols, such as for signaling—SIP, etc.).

- Voice quality

    The goal is to provide the same or even better quality as PSTN, which translated to the Mean Opinion Score (MOS) means > 4.2, only the following codec can provide this quality:

    - G.711—Narrowband audio PSTN
    - G.722—Wideband audio (a.k.a. HD Voice/VoIP)*

    However, in addition to the codec, voice quality also depends very much on the conditions of the IP network where packets are transported. Therefore, to achieve the highest voice quality, the IP network should fulfill the following End-to-End requirements:

    - Delay (mouth-to-ear) < 150 ms for national calls
    - Voice packet loss < 1%
    - Low jitter

    Since the CPE also contributes to these parameters, in order to achieve these requirements, the CPE should have L2 and L3 QoS implemented.

---

* G.722 requires wideband audio terminals.

- Integrated access devices (iADs)

Migration is closely related with broadband coverage as a main technical precondition. As the strategy is to provide voice and other services over broadband access and IP, users need to have an integrated access device (iAD), for example, home gateways (HGW) or optical network termination (ONT), which has to provide interfaces for end user equipment such as telephones, PC, laptop, IP phones, set-up-box, and so forth. At the same time, iADs have a WAN interface as ADSL2+, Optical Fast Ethernet, GPON, and so forth.

Standard ADSL-based HGW configurations typically offer a set of the following interfaces:

- × PSTN FXS ports
- 4 × FE ports
- × WLAN port
- ADSL port

Standard HGW includes firmware, NAT/NPAT, security, QoS (L2/L3), management (TR 069), support DHCP, PPPoE, SIP, VoIP, and so on. They also support several types of voice codices as G.711, G.722, G.729, G.726, and so on.

Optical network termination units, intended to be used with a GPON solution, are equipped with:

- × PSTN FXS ports
- 4 × FE ports

Examples of the above-mentioned simplified schemes are presented in the images in Figure 4.5.

A description of these relationships using a simplified technical diagram will ease presenting the solutions to marketing, sales, and regulatory departments in a very clear and understandable way.

KEY LEARNING

- Use clear and understandable ways for presenting standard migration scenarios to marketing, sales, and regulatory areas.

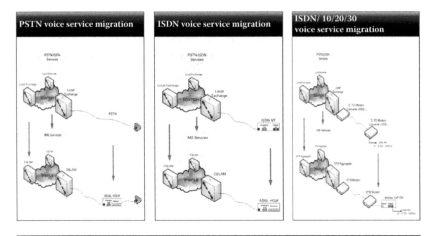

Figure 4.5   End-to-End technical solution examples.

4.1.2.2 Preparing a List of Available Technical IP-Based Features   To start transferring the PSTN products to the IP world, operators first need to prepare the so-called mapping list or correlation matrix of the PSTN features to IP-based available features.

It helps to:

- Gain transparency on IP-based available features
- Provide service continuity to voice customers
- Provide a generic overview and awareness of available technical voice features prior to the start of developing PSTN migration plans by marketing and regulatory departments

4.1.2.2.1 Mapping of PSTN Features to IP-Based Available Features
The preparation process should start at the very beginning, during the deployment phase of the IMS platform.

The following inputs are taken into account:

- A list of current PSTN features taken from the existing product catalogue.
- A list of regulatory obligations in order to be aware of the regulatory and legal service requirements. Due to the fact that most of the features are also required to be provided in the new IP world, the features need to be collected and their implementation has to be ensured.

- Relevant TISPAN, 3GPP, ITU, and ETSI recommendations and standards in order to identify mandatory and recommended voice services according to the relevant technical standards.
- Vendor generic documents for available services in order to identify available technical features in the current software release of the implemented telephony server.

During the preparation process, the main target is the harmonization of the list with marketing, sales, and regulatory areas. At the end, the following outputs are delivered:

- A list of nontransferable services (services to be canceled due to the introduction of IP technology)
- A list of available technical services
- A list of missing services
- New multimedia services

Figure 4.6 presents a checklist for preparing the mapping list; including who to involve and when and which topics will be considered in the mapping of PSTN features versus IP-based available features.

KEY LEARNING

- Marketing and regulatory areas should be made aware of available technical voice features prior to the start of PSTN migration plans.

4.1.2.3 Identification of Differences in Service Ergonomics Due to the Introduction of IP Due to the introduction of new IP technology, some differences in service ergonomics may arise. Therefore, it is important to identify the differences before the migration starts. Differences in service ergonomics mean:

- Improvement of service ergonomics due to an enhanced feature solution implemented on an IMS platform.
- Complexity in service ergonomics due to a CPE limitation.

Topic	What to check	Why is it important	When to check	Responsible
Consideration of complete list of current features/regulatory obligations/product catalogue	• List of current PSTN features • List of regulatory obligations • List of current voice products	• To provide services continuity to the voice customer • To be aware for the regulatory service requirements	• At the very beginning, during deployment phase of the IMS platform	• Technical area in harmonization with marketing and regulatory
Consideration of TISPAN NGN standards/recommendation	• TISPANRec. ETSI TS 181 002; Multimedia Telephony with PSTN/ISDN simulation services • Other relevant TISPAN, 3GPP, ITU and ETSI Recommendations/Standards	• To identify mandatory and recommended voice services according to the relevant technical standards	• At the very beginning, during deployment phase of the IMS platform	• Technical area
Consideration of technical product description doc. provided by the IMS platform vendor	• Vendor generic documents for available services (voice technical product description)	• To identify technical available voice feature in the current software release of telephony server	• At the very beginning, during deployment phase of IMS platform	• Technical area
Consideration of the attached feature list (ANNEX...)	• Mapping list prepared by MKT and HT given in (ANNEX. ...)	• To benefit/take advantage of MKT and HT experience	• At the very beginning, during deployment phase of IMS platform	• Technical area

Figure 4.6 Checklist for preparing a mapping table.

4.1.2.3.1 List of Differences in Service Ergonomics Due to the Introduction of New IP Technology   The following inputs are taken into account:

- Vendor generic documents for available services (voice technical product description) in order to identify the enhancement functionalities.
- Ergonomics of the service in the PSTN world.
- Report from test and verification of the functionalities.

The list of identified differences and a detailed description of the improvement/complexity should be harmonized and approved by marketing before the start of the migration process.

1. Value-added services (VASs) implemented on telephony application server improvements:

   During the process of call setup in a TDM network when the destination end user is in busy status the originating customer receives a busy tone. The improvement is in the generation of the voice announcement "customer busy" in this traffic case.

   Also, voice announcements are introduced into the network during:

   - The activation/deactivation of the services using the special area code (SAC), and
   - During the call setup phase, informing that the call is forwarded.

   Using a network announcement instead of tones during the different traffic cases provides clear information about the call status to the user.

   Also, a benefit of this improvement is the increase of the answer ratio and decrease in offload of the platform from unsuccessful call attempts.

2. CPE dependent limitations:

   Power outage: In case of a power outage at the location of the end customer, the IP-based voice service will be unavailable; service unavailability due to a power outage is an inherent characteristic of IP technology.

A power outage at the location of the end customer should be regarded as a force majeure, which cannot be under the operator's responsibility.

The users need to be informed about this limitation.

Call setup time: With the introduction of the VoIP service, the call setup time is increased.

SAC for activation of the services: During the activation/deactivation of the services using SACs, the standardized SAC must be extended with dialing *# in front. This is the case when connected with AVM CPE devices because of the introduced PBX functionality internal to the CPE device. The benefit of having this internal PBX functionality is the possibility of establishing a call between two users created on the same device dialing a short number.

3. Service platforms:

During the process for the technical product development of the services which could be provided on the existing service platforms, some improvements/limitations were identified as seen in Figure 4.7.

When developing the new IP product portfolio, there are features and terminal equipment which are difficult to transfer. These special cases are split into two groups related to problem identification: Terminal equipment and the potential solution for listed cases are given below.

Terminal equipment:

- PoS terminal
- Alarms/fire detection systems/security systems
- Modem dial-up terminals/remote access

Features:

- Malicious call identification
- Number portability
- Carrier selection (CS) and carrier pre-selection (CPS)
- Wholesale line rental (WLR)
- Extended direct dialing in (EDDI)
- Pulse dialing

	Improvements (+)	Limitations (I)
VAS services implemented on telephony app. server	• Network announcement instead of busy tone • Announcements for activation and deactivation of the services • Call forward service, announcement is introduced in call setup phase "Your call is forwarded"	
CPE related		• Power outage • Increased call setup time • Rotary telephone sets • SAC codes for activation of the services, *# are added in front of SAC for each service in phase of the activation of the service. Case connected with AVM CPE devices
Services implemented on other service platforms	**IVR** • Existing IVR services for fix and mobile users are provided on common platform • New services that are available for mobile subscribers are provided for fixed subscribers as well (exp, Basket Ball, Bingo, Theaters, Cinemas, Flying List, Weather, Thought of the Day)	Due to migration on existing T-mobile platform some features which were not used are not available • For TVT service (Call gapping, in a situation of load of the system on mass TVT event) • AFS (Advanced routing, Announcements, Call gapping)

Figure 4.7   Overview of differences in services ergonomics.

- PBX group
- Public payphones
- CAS PBX

4.1.3 Technical Product Development—Technical Product Design

Additional features that are not supported by the existing deployment of the IMS platform and investments that need to be assessed, and the technical product development process continues with the second step—technical product design.

1. Assessment of current status of network/platform resources:

   In order to identify solutions for services which are not supported by IMS, the first step is to analyze the existing network and to perform proof-of-concept testing.

2. Preparation of the total cost of ownership (TCO) and terminal solution document:

   For introduction of the new services, the technical development process should be followed prior to the handover to operations.

4.1.3.1 Assessment of the Current Status of Network/Platform Resources Based on the list of services which are requested by marketing but not supported by the IMS platform, first an analysis of the potential use of existing fixed and mobile networks is recommended in order to implement the requested service in a shorter and cost-effective way. Terms to be considered include HW configuration, SW configuration, capacity/applications, services traffic, and support contracts.

4.1.3.1.1 Decision Framework for Providing Services on Existing Platforms The preparation of a complete technical and financial analysis (TCO) for platforms in mobile and fixed domains need to take into consideration the existing status, new demands, and a comparison of possible alternatives:

- Consolidation of the existing fixed and mobile platforms for providing services for IMS users.
- Introduction of the new platform integrated with IMS environment.

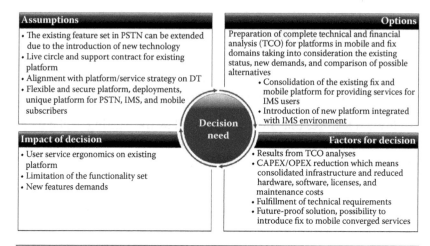

Figure 4.8   Decision framework for providing services on existing platforms.

The process of technical product design for services that can be provided on service platforms in an existing network leads to the following major tasks:

- Comparing the feature set lists in order to identify the limitation or improvements.
- Identifying possible differences in the user service ergonomics.
- Identifying possible regulatory constraints to use the same platform for fixed and mobile network (Figure 4.8).

KEY LEARNING

- CAPEX/OPEX reduction, which means consolidating infrastructure and reducing hardware, software, licenses, and maintenance costs.
- Extension of the service portfolio (in addition to the existing services, new services that are available for mobile subscribers are provided for fixed subscribers as well).
- All platforms (fixed and mobile domains) that need to be reconsidered.

4.1.3.2 Preparation of the TCO and Technical Solution Documents   For the introduction of new features, the TCOs and technical solution documents need to be prepared prior to their implementation.

4.1.3.2.1 Technical Product Development Process Scheme The following activities should be taken into consideration by the local operators for preparing the TCOs:

- An assessment of the current status of the network/platform resources (HW configuration, SW configuration, capacity/ applications, services traffic, support contracts, etc.).
- Alignment with the overall company strategy for the requested services.
- Initiate a proof of concept in order to verify whether the required feature can be provided with the existing resources and minimal investment.
- Investigate the possibility for deployment of the required feature not only in the existing fixed network/platform resources, but also in mobile domains which can provide consolidated infrastructure and reduction of OPEX/CAPEX (e.g., maintenance costs).
- Preparing a TCO analysis and technical solution as a cost basis for determining the total economic value of an investment. From a technical design point of view, it means collecting all the required data for estimating the needed CAPEX/OPEX costs to extend the existing platforms/application servers or implement the new ones.[*]
- Send the technical solution to marketing and depending on the feedback, stop the activities or start with the third step from the TPD process—product implementation.

KEY LEARNING

- Technical product development should be finalized prior to start of the PSTN migration.
- Solutions should be harmonized with all relevant areas in the company.
- Limitations of the service feature set should be clearly presented.
- User guidelines/informative material should be prepared and distributed.
- Proof of concept should be done.

---

[*] Financial information is input for the business case, see Chapter 5.

4.1.4 Technical Product Development—Product Implementation

The third and last step in the technical product development process is product implementation. It starts after receiving positive feedback from marketing, approving the selected features and products, and giving the go ahead for the technical implementation of the requested feature.

This activity is realized in two steps:

1. Start of the technical product development implementation:

   • Identification of needed technical and human resources.
   • Alignment of time plan for technical implementation of the service with the marketing launch plan.
   • Collection of all needed system/interface formats that should be considered in the process of integration of application server/service platform as *input* from billing, provisioning, RAS, fraud, and alarm umbrella.

2. Integration of new platforms or the extension of existing platforms, quality control, and handover:

   • Preparation of handover manuals and guidelines.
   • Preparation of service provisioning and fault clearance procedures, functional requirements for IT development as *output* to IT and operations.
   • Quality control and handover.

4.1.4.1 Start of Technical Product Development Implementation   The process of implementation starts after the technical solution document has been prepared and has harmonized with the marketing area.

In this process, the installation and integration of the platform should be done, including the preparation of the high-level design document as well as low-level design documents.

4.1.4.1.1 List of Resources and Time Allocation for the Technical Implementation of Services   Figure 4.9 provides some information on time and resource requirements for the implementation of different services.

IVR service introduction for IMS users (MKT)

- Decision: Based on the results from the prepared technical and financial analysis (TCO), the migration of the existing IVR applications for fixed users deployed on ULTRA IVR system toward COMVERSE MMIVR System deployed for mobile users was identified as the most feasible solution for hosting IVR applications for PSTN and IMS users
- In-house development of the applications by experts
- Short time to market
- Enhanced feature set

No additional CAPEX and OPEX cost
HR allocation; 120 man days...
Migration of the service within 5 months

Migration of CLASS 4 features on IMS

IMS-based class 4 features, MGC/MGW, SBG, CSCF

CAPEX/OPEX estimation done;
Migration is planned to be performed within 9 months

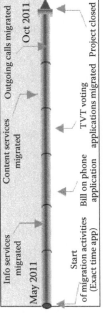

Figure 4.9 Resources and time allocation for the technical implementation of services (examples).

4.1.4.2 Integration of New Platforms or the Extension of Existing Platforms, Quality Control, and Handover   The main activities include:

- Integration with other systems (e.g., access NE, billing, CRM, provisioning, OSS, BSS, existing core network)
- Provisioning of integration guidelines
- NE fields/site deployments
- Preparation of handover manuals and guidelines
- Preparation of service provisioning and fault clearance procedures, functional requirements for IT development as output to IT and operations
- Quality control and handover

The main issues/topics that should be emphasized and considered in the product implementation step are presented in a checklist (in terms of why it is important to check, when, and what to check).

4.1.4.2.1 Checklist during New Service/Platform Introduction   During the process of implementation and integration of the new service/platform the following topics should be considered (Figure 4.10).

During the preparation phase of the technical requirement:

1. Service orchestration:

   During the preparation of the technical requirements for introduction of the new application server integrated on the ISC interface with core IMS, service orchestration should be requested.

   To deliver new application services quickly, complex and changing technology components and systems need to be deployed, arranged, coordinated, and managed.

   Introducing the service orchestration in terms of service architecture provides the possibility of deploying all services on different AS integrated for the IMS core to all users provisioned in the home subscriber service (HSS) node.

2. Limitations identified in one case:

   With the introduction of the business application server integrated with the IMS core platform, supplementary services which were available for some users were not available

Topic	What to check	Why is it important	When to check	Responsible
Introduction of new application server (AS) on ISC interface with Core IMS	• Whether service orchestration is provided? • One user should be possible to trigger all AS integrated on core IMS.	• Not to have limitation of using services on different AS	• During the preparation of the technical requirement for introduction of new AS	• TA
Introduction with Revenue Assurance System (RAS)	• Awareness of the revenue assurance with introduction of the new system or new service • Integration with RAS	• To protect the system from revenue loss due to misconfigured mediation or network nodes	• Prior to or start with final acceptance test • Prior to go live with the service	• TA, IT, and RAS
Intergration with Fraud Management System	• Awareness of the fraud management area with introduction of the new system or new service • Integration with FMS	• To protect the system from unauthorized traffic	• During the implementation • Approval from FMD prior to go live	• TA and FMD
Software firmware on CPE	• Whether the current CPE firmware will support new functionality	• To provide the requested functionality	• At the very beginning during the preparation of the technical requirements	• TA
Service provisioning	• One system for service provisioning on the application server and on core nodes	• Synchronized and on-time provisioning of users and services • Not to have partially provisioned service	• During the implementation phase	• TA, IT
Mediation logic	• What are the relevant nodes for providing event charging information toward mediation device	• To have correct data for billing	• During the implementation phase	• TA, IT, and RAS
Signaling monitoring	• What are all the protocols and interfaces in IMS that network should be considered in implementation	• Monitoring of SIP and RTP traffic in order to be able to troubleshoot the network • Interconnection of ANO operators • Integration of new platform	• Implementation should be done in parallel with introduction of the IMS platform	• TA

Figure 4.10 Checklist during the new service/platform introduction.

for BCS users, for example, call waiting, call transfers, line breaks, anonymous call rejection, and so forth. Marketing expectations were that all value-added services that were available for residential users be available for business users.

3. New firmware on iADs:

During the preparation of the technical solution for the introduction of new services, a mandatory part is to describe the required support of new functionality by the iADs.

With the introduction of the new functionality where the new firmware version of the iAD is needed, the following steps should be performed:

- Define the test list for all services supported by the iAD (voice, Internet, IPTV).
- Perform testing of all services.
- Prepare a plan for remote modification of the firmware version on all types of iADs in use, introducing the new functionality and taking care not to produce long service interruption time.

During the implementation phase:

4. Service provisioning:

In order to have a synchronized, correct on-time provisioning of users and services on all required nodes (not to have partially provisioned users), service provisioning should be controlled by one centralized system having feedback information about the status of the performed provisioning activity (successful/unsuccessful with reason).

5. Mediation:

During the implementation phase, the relevant nodes which should provide the event charging information to the mediation device for each traffic case and service should be identified. The operator's experts should participate in the creation of the correlation logic on the mediation device. CDRs per each traffic case and service should be checked during the quality control.

6. Integration with revenue assurance system (RAS):

With introduction of the new service/platform in order to assure revenue, integration with the RAS is mandatory. The following activities should be performed:

- Define two sources for loading data to the RAS system (mediation CDRs and signaling monitoring CDRs)
- Definition of the list of traffic cases relevant for RAS
- Definition of the correlation logic in signaling monitoring system in order not to have duplicated CDRs

7. Integration with fraud management system (FMS):

With the introduction of the new service/platform to protect the system from unauthorized traffic, the integration with FMS is mandatory. The following activities should be performed:

- Define two sources for loading data into the FMS system (mediation CDRs and signaling monitoring CDRs)
- Definition of the list of traffic cases relevant for FMS
- Identify business users that can generate huge traffic
- Identify the required parameters in the CDRs (IP address, VLAN ID, normality flag, etc.)

8. Signaling monitoring system:

Introduction of the signaling monitoring system is recommended to be performed during the implementation phase of the IMS platform.

A table should be prepared in order to identify the mandatory protocols and interfaces to be monitored. Also the monitoring points must be defined.

Implementation of the signaling monitoring system helps during the troubleshooting process of the service functionality and network performance, during the process of interconnection with alternative operators, and during the process of the introduction of the new service/service platform.

KEY LEARNING

- Service orchestration should be requested from the vendor.
- Integration of the new AS with the core platform is a very complex project and requires a significant amount of time for implementation.
- Additional licenses are needed for core nodes.
- Synchronized and on-time provisioning should be provided.
- Mediation logic (event charging information from AS and core nodes).

4.2 NT Rollout

The NT rollout has two elements:

I. Network and technical product readiness

- Planning the adjustments of the network elements.

II. Internal resources preparation and planning

- Planning the actual migration support by technicians as well as the training of the technicians (for training, see also Section 6.3).

4.2.1 Network and Technical Product Readiness

Network and technical product readiness has two subactivities:

1. Definition of migration enablers

- Identifying the necessary adaptations in the network.

2. Development of network deployment plans

- Planning network adjustments.

4.2.1.1 Definition of the Migration Enabler The definition of the necessary enablers in the network requires a thorough analysis of the network. These can comprise elements in core, service platforms, access network, iADs, and supporting systems.

4.2.1.1.1 List of the Impacted Network Layer Elements This element lists all necessary changes in migration enablers/network elements that should be implemented in order to introduce and to start the customer migration process.

As inputs are needed:

- List of network elements in usage by the particular operator with defined links among them.
- Technical product development plan from Section 4.1.2 of this document.
- Migration concept and time plan from the migration plan.

IT is divided into core and access elements with the core part describing platforms and the access part listing last mile equipment.

The first prerequisite, from a core point of view, is the IMS platform. Also, changes in IP transport are necessary, while all other platforms are "nice to have." For access, adequate terminal equipment (IMS ready) is essential as well as the access port (Figure 4.11).

KEY LEARNING

- Changes have to be implemented on time.
- Any delay prevents the ultimate goal—old technology switch-off.

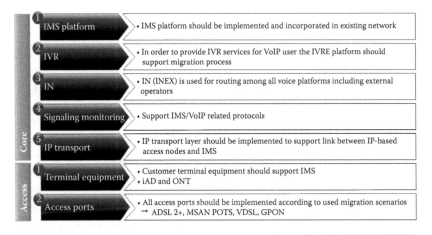

Figure 4.11 List of impacted network layer elements.

4.2.1.2 Major Network Deployment Plans   Preparation of deployment plans must be aligned with:

- Overall network architecture
- Migration plan
- Decisions from technical development plans
- Detailed customer services migration plans
- Unrelated migration connection\sales plans received from marketing and sales

There are high-level plans that reflect all network structure and elements and low-level plans that reflect access network and CPE procurement as an essential preparation part for the PSTN migration.

4.2.1.2.1 Examples of the NT Rollout Plans and Learning   For the preparation of deployment plans for the application level, transport level, access, and CPE level the following inputs must be provided as a technical precondition and cost per equipment:

Application level, service platform roadmap which should include information about:

- Live time of the platforms and support contract (expiration date and support cost)
- Obsolete platform in terms of existing platform that cannot provide the services for IMS users
- Timeline for implementation of the services
- Target network architecture
- Cost per user/service license

Transport network:

- Existing capacity and necessity of capacity extension regarding PSTN migration and connection/sales plan.

Access network (narrowband and broadband access):

- Broadband penetration and coverage. PSTN migration cannot be isolated from the broadband transformation since users shall have access to all services (including voice) over one common broadband access. Therefore, broadband penetration is one of the main prerequisites for voice migration.

- Customer access line requirements (capability, minimum upstream and downstream bandwidth).
- Existing and planned capacity
- Limitations
- Use cases, solutions for overcoming problems
- Cost per access port

CPE devices:

- Type of the CPE device per service per migration scenario, and
- Cost per CPE device.

Core development plan:

- The high-level plan should include reference to all changes that IMS implementation and PSTN migration implicate on plan preparation.

Access deployment plan:

- Low-level plans for access network deployment should be detailed in terms of capacities and time schedules.

The main inputs for analysis and decision factors are:

- PSTN migration plan and connection/sales plan—access network deployment to be in line with detailed customer migration plan.
- CAPEX plan and prices for quantities versus warranty of equipment. The warranty of equipment starts from the delivery date and it is not justified to build stock of unused equipment unless the discount for quantities is essential.

CPE procurement plan:

- Like the access network rollout plan, the CPE procurement plan is related to the PSTN migration plan, connection/sales plan, and quantities. An additional parameter is the spare part stock that must be considered for CPEs.

The main inputs for analysis and decision factors are:

- Quantities in stock
- Expected price decrease/increase

- Expected discount for large quantity procurement
- The very important factor for procurement decisions is the level of readiness of HGW types regarding the IMS features and importance of the problems which are identified. If the problems and deficiencies are essential, than the vendors must be pressed to make final changes. The procurement of bigger quantities should be after testing of a few samples.

KEY LEARNING

- Testing must be performed on all services (not only IMS related) because with every new FW, problems with IPTV or Internet service were identified which were not expected.
- For monitoring IMS readiness, regular reports should be prepared.
- Plans for access network deployment should be adjusted to regional PSTN migration plans.
- Plan for CPE deployment should be in relation with other non-PSTN migration projects (connection plan received from sales).

4.2.2 Internal Resources Preparation and Planning

1. Customer premises rollout plan: Definition of customer premises rollout plan as a detailed planning of technical service.
2. Internal resources preparation—training:

   - Prepare detailed and different training approaches
   - Perform training and request feedback
   - Adjust training according to key learning and best practice

3. Upsell technician as salesman: Use actual project as potential for upsales.

The planning and preparation of the technical workforce is an essential activity and covered with all other training in the training chapter of "Go to Market" (see Chapter 6, Section 6.3.2).

4.3 IT Roadmap

The blueprint of the IT roadmap describes the necessary changes and developments that have to be done in order to prepare the IT systems to support the IMS implementation and to provide smooth provisioning processes for IP products and the PSTN migration process. Two main deliverables are identified that are important in this IT systems preparation.

I. IT application and product readiness:

Identification of changes in IT systems and interfaces, development, and implementation in order to integrate the IMS system with BSS and OSS systems in the company.

II. Customer migration process within IT:

Development of IT systems for supporting PSTN migration processes and efficient provisioning of IP-based products/services.

4.3.1 IT Application and Product Readiness

This element relates to the implementation of IMS into the IT architecture, adopting the IT systems for supporting new IMS products/services, and integrating IMS with the IT systems from BSS and OSS domains (CRM, billing, OSS, and DWH/BI).

There are three major activities for this deliverable:

1. Identification of required changes in IT applications.
2. Adaptations of existing IT systems/development of new IT systems.
3. Adaptation of interfaces toward the network.

4.3.1.1 Identification of the Required Changes to IT Applications Transfer from the PSTN world into the IP world is a significant change for telco providers and requires adaptation of all affected IT systems in order to provide smooth implementation of the IMS platform and services supported by this platform. For this purpose, the process of identification of the changes and affected IT systems is the first step for IMS implementation to prepare the technical requirements for the system. This activity is focused on the identification and specification of changes and functionalities that have to be implemented in the IT systems.

Major subactivities in this project phase are:

1. Mapping of marketing requirements with the technical requirements and identification of gaps.
2. Identification of changes that should be made in IT applications in order to support the PSTN migration process to IMS.
3. Identification of changes that should be made to interfaces for integration with the involved system in the PSTN process.

Based on the marketing and technical requirements, IT should prepare the analyses for identification of the changes and gaps in the IT systems and interfaces, and consider:

- Existing IT systems
- Current IT systems architecture
- Implemented product structure

This activity is important to assure that all products will be implemented as required and to find a solution for products that are not natively supported by the existing systems/platforms.

Awareness for this process is not only required in NT and IT areas but for marketing and sales as well, in order to be able to define the product strategy for the IP world and to be able to prepare the marketing requirements for the IMS implementation. On the other side, the technical departments have to prepare the technical requirements to align with the capabilities of the new IMS platform and define the provisioning process for the new products/services. During the process of preparation of marketing and technical requirements toward IT, it should be considered that there are some services that cannot be provisioned due to some limitations by the IMS platform, corresponding to regulatory or CPE issues (fax, POS, ATM), and so forth. All these cases should be identified and the requirements have to provide the information for how those services will be provided in the future.

The activity yields identified high-level changes in the systems and in IT architecture including needed resources and a time plan.

4.3.1.1.1 Checklist of High-Level Changes That Have to Be Implemented in the IT Systems This section provides an indication of potentially required high-level changes that have to be made for implementing the IMS as well as the new products and features. There are several aspects that must be taken into account:

- Capabilities of existing systems—They should be checked.
  - Does the existing IT system support all requirements or should new systems be implemented?
  - There should be analyses of levels of changes in the existing systems, and the effort needed regarding the resources and budget, compared with replacement with the new system.
  - The possibility for reuse of some parts of the existing solutions for voice services should be considered.
- Capabilities of existing interfaces—They should be checked to see whether to use the native IMS interfaces or develop an intermediate level for interfaces.
- Capabilities of mediation system for processing IMS records.

For the assessment, experts from all affected IT domains should be involved in the analysis. According to experience in the countries, the affected domains are:

- Billing domain—Changes in the product structure and mediation system.
- CRM—Changes in the product structure and interfaces with the service/resource domain.
- Service/resource domain—Changes in the inventory systems and provisioning processes.
- DWH/BI—Adjustment of the data model, ETL processes, and development of the reports.

The final output of the analyses should result in a:

- High-level design of IT systems
- List of interfaces needed to support the IMS implementation
- New IT architecture
- Specifications for the cost and time implementation

Figure 4.12 shows important checks that should be performed before starting with other activities regarding PSTN migration.

According to the countries' experience, identified functionalities in IT applications and integration platforms are stated below.

- Integration platform:
  - IMS interface for provisioning
  - Interface to DSLAM/GPON/MSAN
  - Interface to switching
  - Adjust integration with CPE management system to support IMS information
  - CRM interfaces to OSS, billing, and DWH
  - OSS interfaces (northbound and southbound)
- Application:
  - Adapt/implement system to use IMS interface for provisioning of services
  - Change/implement workflow/processes in order to apply IMS logic
  - Change/implement data structure in technical inventory in order to store additional information needed for IMS
  - Change/implement applications GUI in order to manage additional information needed for IMS
  - Ensure solutions which include old voice services
  - Implement new products in BSS/OSS for IMS
- Adjust product decomposition from CRM in order to support IMS.

KEY LEARNING

- Reuse existing logic for interfaces if possible in the new technical design.
- Reuse the existing process for voice provisioning if possible.
- Define the final approach and do not change it.

4.3.1.2 Adaptation of the Existing IT Systems/Development of New IT Systems
This activity uses the specification of the changes and high-level design (output in Section 4.3.1.1) to identify the exact systems that

Topic	What to check	Why is it important	When to check
Mapping of marketing and technical requirements and identification of gaps	• Are all existing products supported by IMS • Is marketing planning to use the same products for IMS or will they create new ones	• To find the solution for nonsupported products • To plan the implementation of the products in the support systems	• When both requirements will be available, but in the analyses phase of IMS implementation project
Identification of changes that should be made in IT applications in order to support PSTN migration process to IMS	• Capabilities of mediation system for processing IMS records • Possibility for reuse of some parts of existing solutions for voice service • Find the way to support old voice services (PSTN/ISDN)	• To decide whether to use the existing mediation or to invest in the new one • Decrease development effort • In order to support all types of different customer equipment that natively might not work on IMS	• Analyses phase and preparation of technical requirements for IMS
Identification of changes that should be made to interfaces for integration with involved system in PSTN process	• Check whether to use native IMS interfaces or develop intermediate level for interfaces • Integration with CPE management system to support IMS data	• To define the strategy which is less time/budget consuming • To see if something in the CPE management system has to be updated • To provide efficient provisioning process	• Analyses and design phase of IMS implementation project

Figure 4.12 Checklist of high-level changes that have to be implemented in the IT systems.

have to be changed or new systems that have to be developed. Based on that assessment, the implementation of changes is initiated and executed, yielding an all-IP-ready IT infrastructure.

The following subactivities have to be performed:

1. Analysis of the functionality of the existing systems and possibility of adjustment
2. Decision for the adaptation of an existing or the implementation of a new system
3. Development of IT applications

4.3.1.2.1 List of the Applications/Systems to Be Adjusted or Developed with Potential Issues and Solutions Based on identified changes with the high-level design, changes in the applications should be made. A detailed process specification and product definition are essential and must be ready with all specifications. Changes in applications will probably affect all of the following systems (Figure 4.13).

- Application in OSS for provisioning of services:
  - Extension of inventory system with data needed for IMS—with the IMS there might be the need for some new data in inventory. Examples are phone context, password for voice account, password for Web page for self-support; if inventory is able to store this information, but if it cannot then extraction is needed.

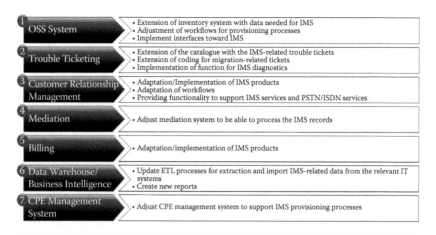

Figure 4.13 List of applications/systems to be adjusted or developed.

- Adjustment of workflows for provisioning processes—The system for workflow must be configured/developed for supporting the new process for IMS provisioning. The process of provisioning will depend on the technical solution that you make, but it is most likely that it will be different from the PSTN provisioning process.
- Implement interfaces toward IMS—This is something that must be done. IMS interfaces for provisioning must be implemented in the OSS system that is processing service orders for voice. If there is VoIP provisioning to be done on an existing platform, consider reusing the existing logic implemented in the application if possible.
- Trouble ticketing:
  - Extension of the catalogue with the IMS-related trouble ticket—Here a configuration/development of a new type of trouble ticket should be done, because solving tickets for voice on PSTN is different than those on IMS. Different possible error types should be defined. Depending on the organization, maybe different departments will solve those issues. Also, presented information in the ticket may need to be extended with the new data related for IMS.
  - Implementation of the function for the IMS diagnostic—It is recommended to implement some diagnostic in the trouble ticketing system related to IMS. At least the status of a phone on IMS such as not created, registered, or not registered, should be available to the trouble ticketing system. Depending on the solution, for the order service to be functional, some specific information may be necessary, such as the customer locations function (CLF) information for location of customer on DSLAM/OLT.
- Customer relationship management:
  - Adaptation/implementation of IMS-related products—This must be done and products must be configured in the CRM application.
  - Adaptation of workflows—Once the workflow in CRM for the new/adapted products is different, the adaptation of this workflow needs to be taken into account for the time plan.

- Providing functionality to support IP-based services and old voice services—Do not support only IP-based voice service in the CRM. There are cases (wholesale, special devices on the customer side, etc.) of when to support PSTN voice in CRM. Plan to provide this from the beginning.
- Mediation:
  - Adjust the mediation system to be able to process the IMS record—IMS records are different than records that are received from PSTN switches. Changes in mediation must be done and depends on the system that is purchased from the IMS vendor.
- Billing:
  - Adaptation/implementation of IP-based products—Implementing IP-based products for billing may just be a configuration of new products, but this depends on the billing system and on the design of products from marketing.
- Data warehouse/business intelligence:
  - Update "extract, transform, load" (ETL) processes for extraction and import IMS-related data from the relevant IT systems and IMS platform—implementing new data in the inventory system will produce this change in the DWH system.
  - Create new reports—For any future migration process, having reports is essential. The needed reports must be specified and well defined before starting any development in systems.
- CPE management system:
  - Adjust the CPE management system to support IMS provisioning/fault repair processes—Modification of parameters that are provisioned should be done for voice service. Those parameters may depend on the type of CPE.
- IMS user interface:
  - Out of the box there is no easy way to read complete service data directly from an IMS on a provided GUI from the vendor. The data should be read from many IMS nodes/systems to obtain simple information such as the status of the services for one customer. Also, applying

new settings for voice services is not easily done from the provided GUI from the vendor. All service changes are done through XML scripts that should be populated and executed manually (possibility for errors). That is why the tool for reading/modifying data in the IMS for customers is much needed and should be considered to be developed or bought.

4.3.1.3 Adaptation of Interfaces toward the Network In order to provide interoperability and smooth integration of IMS platforms into the IT/NT architecture, it is necessary to adapt or develop interfaces between the systems. Inputs for this activity are documented for platform, list of affected interfaces, process specification, and IT architecture and this activity will result in updated or newly developed interfaces.

The following subactivities have to be performed:

1. Identification of interfaces that have to be modified and new interfaces to be developed
2. Mapping with NT/IT architecture
3. Technical specification of interfaces
4. Development of interfaces

4.3.1.3.1 List of the Mappings and Interfaces That Have to Be Modified/ Developed Based on identified changes with the high-level design and the list of interfaces, changes should be made in the existing integrations and there should possibly be development of new interfaces. Detailed process specification and vendor documentation for interfaces are essential and must be ready with all particulars.

High-level mappings and interfaces that have to be modified/ developed are presented in Figure 4.14.

4.3.1.3.1.1 IMS Interface for Provisioning The aim of the IMS is to provide voice services on an IMS platform. To start, define what should be executed on IMS in order to create voice service. Additionally, for all VAS that should be provisioned on IMS, all parameters of the XML scripts must be available. These scripts can differ between telecommunications operators, especially in part of the barring programs and regulatory services. A sample mapping is shown in Table 4.1.

		IT Domain			
		CRM	**Billing**	**OSS**	**Data warehouse**
Interfaces	IMS interface for provisioning	✓	n/a	✓	n/a
	Interface to DSLAM/GPON	n/a	n/a	✓	n/a
	Interface to switching	n/a	n/a	✓	n/a
	Adjust integration with CPE mgmt. system	n/a	n/a	✓	n/a
	CRM interfaces	✓	✓	✓	✓
	OSS interfaces	✓	n/a	✓	✓

Figure 4.14   List of mappings and interfaces to be adjusted or developed.

Table 4.1   Sample Mapping

IDVAS	SERVICE	ACTION	XML SCRIPT NAME
1	CLIP	Activate	Originating identity presentation, activate
1	CLIP	Delete	Originating identity presentation, delete
2	CLIR	Activate	Originating identity presentation restriction, activate
2	CLIR	Delete	Originating identity presentation restriction, delete
7	Conference call	Activate	Conference, activate
7	Conference call	Delete	Conference, delete

This interface is mainly used by the OSS domain (part for the provisioning), but in some cases this interface can be used from the CRM domain (self-service, diagnostic).

- If an existing interface to other VoIP platform is available, reuse the existing logic for provisioning interfaces.
- Depending on the provisioning system there is a possibility for direct usage of provided provisioning interfaces or to build a transformation layer between the provisioning system and IMS interfaces. Analyze the faster and more flexible solution:
  - Many changes/adaptation in internal systems, or
  - Build new layer, which requires minor changes only in internal systems.

Interfaces to DSLAM/GPON/MSAN provisioning—interfaces that should provide automatic provisioning of access service for all technologies. They are not directly connected to IMS, but are very important for the whole provisioning process for IP-based services. The interfaces are used from the OSS domain. If there is an integration to DSLAM/GPON/MSAN then there will be a need to make adaptations.

- Interface to switching/call routings—This interface should be used during the IMS provisioning process to generate necessary changes in the PSTN switches/IN platform in order to provide proper functioning of all types of customer calls (in or out) from-to all voice platform. This interface is used from the OSS domain (part for the provisioning).
    - Depending on the solution for call flows through the voice platforms, will lead to the next steps for voice service on the switch. This, for example, can be complete removal or making a route to IMS.
    - The availability of this interface will speed the provisioning time and lower the amount of trouble tickets that will arise from the migration process.
- Adjust integration with the CPE management system—This interface is used for zero touch provisioning (to send firmware, voice configuration data) to the CPE device. Depending on the actual device types you may need to make different configurations in order to provide voice on IMS. This interface is used from the OSS domain (part for the provisioning).
    - Having this interface will speed the provisioning time and lower the amount of trouble tickets arising from the migration process.
- CRM interfaces—The interfaces from CRM to other domains may need to be changed due to the PSTN migration project.
    - To the CRM domain—Adaptation of the interfaces that are used from some systems in the CRM domain, for example, self-service portal (new type of requests/products should be configured). Data transfer: CRM → CRM.
    - To the billing domain—Extension of the interface from CRM to billing if CRM products depend on technology. Data transfer: CRM → billing.
    - To the OSS domain—Extension of the interface for orders from CRM to OSS with the IMS data or mappings. Data transfer: CRM → OSS.

- To the DWH—Extension of the interface for data collection from DWH regarding new IMS data that will be required in the reports. Data transfer: CRM → DWH.
- OSS interfaces—Those interfaces from OSS to other domains may need to be changed due to the PSTN migration project.
  - To CRM—Extension of the interface for order processing with IMS data or mappings. If some diagnostic is provided to the call center, then extension of that interface should also be made. Data transfer: OSS → CRM.
  - To the OSS domain—Adaptation of the interfaces that are used from some systems in the OSS domain, for example, trouble ticketing (extension of ticket enrichment with IMS data). Data transfer: OSS → OSS.
  - To the DWH—Extension of the interface for data collection from the DWH regarding new IMS data that will be required in the reports. Data transfer: OSS → DWH.

KEY LEARNING

- Build the transformation layer between the IMS and the internal system.
- Test interfaces into implemented applications with the E2E process before going into production.
- Some customer equipment may not work properly on IMS. Build an interface for creating voice on old platforms (PSTN/ISDN).
- Consider the utilization of an existing mediation system.

4.3.2 Customer Migration Process within IT

This element ensures readiness of processes for migration, which includes:

1. Development of migration process
2. Development/adjustment of all "regular" processes for voice services
3. Enabling of migration reporting

4.3.2.1 Development of the Migration Process One of the basic conditions for the VoIP migration is the specification, development, and implementation of processes for migration and processes for VoIP provisioning and fault repair. It is also important to define all the dependencies of the migration process with other business processes to facilitate implementation of user requirements.

Although demanded changes in IT systems are difficult, it is important to ensure the possibility of realization of customer requirements during VoIP migration in the implementation phase. The preparation and implementation of the required changes in IT systems are essential to ensure the normal functioning of provisioning and fault repair processes for all requests by customers migrated to VoIP. This is very important due to the complex processes that include retail and wholesale requirements.

Preparation of the technical specifications requires detailed and clear functional specifications from all stakeholders involved in the migration process and all affected business processes. Functional specifications must include a clear description of the processes (graphical representation), business rules, responsibilities, input and output data, the contact points between the different business units, the data exchanged between the systems involved in supporting the realization of the process, and the definition of data required for reports and analytics. Technical specifications should cover clearly defined necessary changes in the IT system as well as in the E2E environment.

It is also essential to prepare the test book for acceptance testing of required changes. This test book should be used in the integration testing phase to ensure the highest quality of testing developed processes and thus reducing the time for the duration of acceptance testing and reducing the number of errors.

Major activities in this project phase are:

- Analysis of submitted requests for changes in IT systems.
- Preparation of technical specifications based on functional specifications for new processes or changes to existing processes.
- Implementation of changes in IT systems with detailed testing.

Based on detailed functional, marketing, and technical service specifications IT should analyze all requirements and prepare IT technical specifications for implementation changes in support systems.

During the analysis, it is required to take into account the following restrictions and interdependencies:

- Limitation of current IT architecture.
- Available resources.
- Interdependence with other projects and regulatory requirements.

The final output of the development of migration processes should result with the:

- Technical specification for changes in IT systems.
- Results of integration and acceptance testing.
- Manuals for work in IT systems.
- Implementation of changes in IT systems.

4.3.2.1.1 Checklist of Changes That Have to Be Implemented in the IT Systems to Support the Migration Processes   Figure 4.15 shows some important checks that should be done before starting with other activities in the PSTN migration.

For the activity, "Analysis of submitted requests for changes in IT systems," it is important to check if existing processes support change requests related to PSTN migration and if functional specifications contain all the elements needed to create the technical specifications. These checks will allow optimal implementation of the required changes to the existing IT systems.

In the activity, "Preparation of technical specifications based on functional specifications for new processes or changes to existing processes," check if there is the possibility for the reuse of parts of existing implemented flows for business processes, and whether the technical specification includes all business rules. These checks may decrease the development effort in order to support all business requirements and reduce the number of errors in the later stages of development.

In the activity, "Implementation of changes in IT systems with detailed testing," check if developed functionality is correlated with technical specification. Decrease the number of errors during the acceptance test and implementation phase. This contributes significantly to maintaining the quality of delivery services to customers who have migrated to VoIP.

Topic	What to check	Why is it important	When to check
Analysis of submitted requests for changes in IT systems	• Do existing processes support change requests related to IMS migration • Do the functional specifications contain all the elements needed to create the technical specifications	• To find the solution for non-supported processes • To plan the implementation of the processes in the supportive systems	• When the business requirements will be available, but in the analyses phase of preparation of technical specifications
Preparation of techn. specifications based on functional specifications for new processes or changes to existing processes	• Possibility for reuse of some parts of existing implemented flows for business processes • Whether the technical specification includes all business rules	• To decide whether to use the existing flows or to develop the new one • Decrease the development effort in order to support all business requirements • Reducing the number of errors in the later stages of development	• During preparation of technical specifications
Implementation of changes in IT systems with detailed testing	• Checking the developed functionality with technical specification	• Reducing the number of errors during the acceptance test and implementation	• During integration test

Figure 4.15 Checklist of changes that have to be implemented in the IT systems to support migration processes.

4.3.2.1.1.1 List of Migration Processes to Be Developed   The list in Table 4.2 states the migration processes to be developed and integrated, categorized into priority 1 (business process which should be implemented at the start of the migration process) and priority 2 (business process which should be implemented after the start of the migration process as early as possible).

Table 4.2   List of Processes to Be Developed

PROCESSES	PRIORITY
Migration POTS→VoIP (MSAN); only for solution with 1 number on IMS; NatCom installation	1
Migration POTS→VoIP (ADSL or Optics); user has appropriate modem; automigration	1
Migration POTS→VoIP (ADSL or Optics); user has no appropriate modem; self-migration	1
Migration ISDN BRA→VoIP (ADSL or Optics); user has no BB services; NatCom installation; solution with 2 or 3 numbers on IMS	1
Migration ISDN BRA→VoIP (ADSL or Optics); user has appropriate modem; automigration; solution with 2 or 3 numbers on IMS	1
Migration ISDN BRA→VoIP (ADSL or Optics); user has no appropriate modem; self-migration; solution with 2 or 3 numbers on IMS	1
Migration started from technician (technician like salesman)	1
VoIP connection (MSAN)	1
VoIP connection (ADSL or Optics); user already has BB (broadband) services or not; solution with 2 or 3 numbers	1
Change on VoIP (additional services or number change); MSAN, ADSL, or Optics	1
Temporary disconnection of VoIP (and BB services); MSAN, ADSL, or Optics	1
VoIP disconnection (and BB services); MSAN, ADSL, or Optics	1
VoIP+NPP processes (NPP-Netphone package; VoIP service for business customer on different platform)	2
VoIP connection for user who migrates from another operator (MSAN, ADSL, or Optics)	2
VoIP relocation (and BB); ADSL→ADSL, POTS, ADSL→ADSL; ISDN BRA, ADSL→ADSL; MSAN→MSAN	1
VoIP relocation (and BB); POTS→MSAN; MSAN→POTS; Optics→Optics; POTS, Optics→Optics; ISDN, Optics→Optics	2
VoIP disconnection+WS (wholesales) order (NP [number portability], ULL [unbanding local loop]; BSA [bitstream]); MSAN, ADSL, or Optics	2
VoIP +NPP disconnection+WS (wholesales) order (NP [number portability], ULL [unbanding local loop]; BSA [bitstream]); MSAN, ADSL, or Optics	2
Migration MSAN→ADSL (connection of solution with 2 or 3 number and/or BB service)	2
Migration MSAN→Optics (connection of solution with 2 or 3 number and/or BB service)	2
Migration ADSL→MSAN (BSA connection)	1
Migration ADSL→POTS (BSA connection)	1
VoIP fault repair (MSAN, ADSL, or Optics)	1

KEY LEARNING

- Analyze all the processes necessary for migration, provisioning, and fault repair in scenarios for sales and wholesale.
- Make the prioritization process for the implementation of support systems.
- Allow user requests for services during migration.
- Introduce more than one channel for migration (example "2 to 1" migration case) at the beginning of the migration project.
- Ensure and test quick system response while the technician is on the customer site.
- Good organization and conduct testing are crucial.

4.3.2.2 Development of All Regular Processes In addition to the processes for supporting migration itself, it is necessary to define and implement all related processes that will support business as usual (BAU) during and after PSTN migration. This activity uses defined priorities for the processes, business plan, and functional specifications for all related processes in order to provide a detailed technical solution and implementation of those processes in IT systems.

In a first step, analysis is performed to define the priority of processes introduction, followed by the implementation of changes in IT systems according to the priority.

4.3.2.2.1 Prioritization of the Logic Processes and Learning List In order to accomplish successful migration to the new platform, in addition to the definition and implementation of migration processes, it is crucial to define all the other necessary processes involving the new platform (business as usual). Since the new platform brings not only new interfaces, but also new services and therefore necessary changes in product design, it is important to implement all of the changes in IT systems. Also, parallel with migration, business goes on with the development of new products, new offers, and all other changes, which also need implementation within IT systems. It is likely that not all the changes can be made instantly or at the same time, however, they can be prioritized. In order to make optimal prioritization, a few aspects must be taken into consideration.

- Category of process change:
    - Basic processes for migration itself
    - Basic processes for provisioning on a new platform
    - Service change processes
    - Business driven new processes
    - Complex processes on already provisioned services (e.g., relocation, takeover, etc.)
    - Regulatory driven processes (ULL, bit stream, etc.)
- Business plans for the period of migration.
- Available resources for implementation of changes.

Since IT cannot prioritize the changes by itself, the best way to do this is to form a formal organization (committee) involving all stakeholders from inside the company with the power to make decisions on priority and resource assignments for each defined change. This committee will prepare priorities and together with functional specification a business plan that will be input for rollout of all necessary process changes in IT systems. Based on the experience, the list of process prioritization is given in Table 4.2, "List of Processes" in Section 4.3.2.1.

Awareness of the whole PSTN migration project must be present at each part of the organization inside the company and interdependencies with other projects must be minimized. Also, change requests in business as usual must be aligned with resource planning of the PSTN migration project, because too many changes in IT systems can jeopardize migration and system readiness after successful migration.

It is obvious that, after development of migration processes itself, the first priority is to introduce basic processes for provisioning on the new platform and leave old PSTN provision processes behind as much as possible. After that, it is very important to introduce processes to support regulatory requests, which are obligatory by law. In case of a stepwise approach, processes from later phases have to be supported by workaround using the existing processes.

Topic	What to check	Why is it important	When to check
Analysis and definition of the priority of processes introducing	• Check the importance of the process in relation to the defined criteria (frequency, importance, migration plan)	• Covering the most important and common business processes during and after migration	• Before preparation of technical specifications
Implementation of changes in IT systems according to the priority	• No need for additional checks related to prioritization because everything is defined through technical specification	• Implementation according to defined priorities	• During the implementation phase

Figure 4.16 Prioritization logic processes and learning list.

In Figure 4.16, there are some important checks that should be done before all processes are prioritized and put into plan for implementation.

In the activity in Figure 4.16, "Analysis and definition of the priority of processes introducing," check the importance of the process in relation to the defined criteria (frequency, importance, and migration plan). These checks ensure covering common business processes prior to any other during and after migration. All of these checks should be made during the preparation of the functional specifications and business requirements prior to the preparation of the technical specifications.

In the activity, "Implementation of changes in IT systems according to the priority," there is no need for additional checks of prioritization because all of the checks need to be done at the first step and everything is defined through the technical specifications. This activity should follow the prioritization plan and ensure deliveries according to that plan on time and within planned budget. These activities are done during the implementation phase itself.

KEY LEARNING

- Prioritization should be based according to the importance and frequency of the process.
- Prioritization should be based according to migration plans.
- Automation should have a high priority with the consideration of the importance and frequency of the process.

4.3.2.3 Enable Migration Reporting   Migration monitoring and reporting are essential components of migration management and crucial for overall success and performance. It should be used among others to ensure that migration processes are implemented as efficiently and effectively as possible.

Reporting is essential for the continuous process of assessing the status of migration in relation to the approved work plan and budget. It helps to improve performance and achieve results. The overall purpose is to ensure effectively managed results and outputs through measurement and assessment of performance. It is particularly important in the design of a realistic chain of results, outcomes, outputs, and activities. Regular monitoring enables the possibility to identify actual or potential problems as early as possible in order to facilitate timely adjustments in migration implementation.

Together with changes in processes and systems it is very important to ensure efficient reports for monitoring, management, and support on an operational level. Therefore, all relevant data has to be collected from source systems into data warehouse (DWH) and developed into the readable reporting for end users.

The following subactivities are recommended:

- Analysis of submitted report requests by the business side versus already existing reports in the DWH system.
- Preparation of the DWH system and end systems (BSS/OSS) for new reports.
- Implementation of new reports in DWH.

4.3.2.3.1 List of the Data Sources and Reports Including Reporting Users Project reporting is the formal presentation of monitoring information. The main reasons for reporting include the following:

- *To formally inform management*: Reporting ensures that management is formally apprised of the progress made in migration and aware at an early stage of actual and potential problems and any remedial action taken.
- *To help migration planning and execution*: Reporting ensures continuous and focused migration planning and execution being in principle a long-term project.

- *To serve as an audit*: Reporting maintains a record of all actions taken. It therefore constitutes a vital resource for performance analysis and helps in the identification of irregularities as well as creation of corrective measures.
- *Operations reports*: In absolute numbers, these are the most numerous reports, embedded in all standard process.

There are a few things that should not be forgotten—good reporting and monitoring should be in line with:

- *Focus on results and follow-ups*: Look for "what is going well" and "what is not progressing" in terms of progress toward the intended results.
- *Regular assessment by the migration team*: The resource should be dedicated to assessing progress, looking at the big picture, and analyzing problem areas.
- *Regular analysis of reports*: The team should continuously review migration project-related reports.

In the following enumeration, there are some common data sources engaged in IP transformation standard reporting. Keep in mind that this list will vary based on the status and implementation of systems in OSS/BSS landscape.

Data sources:

- CRM—customer relationship management—customer data
- Inventory system
- Trouble ticketing
- WWMS—workflow and workforce management system
- Billing system
- CPE management
- IMS platform data—mainly HSS data

List of reports, short description, and report main users:

- Orders and applications count

  This report shows the current situation of open/closed work orders from the CRM system and connected open/closed work orders from the service provisioning (SP) system. It also gives the current situation of open SP work orders grouped by departments.

- Orders by type—last X days

    This report shows open/closed SP work orders by type and service in last X days (where X is the prompt value). This is a daily report.

- List of orders waiting more than 5 days

    This is a daily report representing a detailed list of all SP work orders that are still open in the system for more than 5 days.

- Network analysis

    Gives snapshot of current line situation in the whole network. It is grouped by geography and switch information. It represents the number of lines with or without certain services, equipment, or platform (in case of IMS).

- Lines on virtual DSLAM

    This is a daily snapshot report with current virtual DSLAM information grouped by switch and service information.

- Nonmigrated customers during service provisioning

    Weekly summary of work orders for nonmigrated customers from SP grouped by reason code.

- Call center PSTN migration report

    This is a daily report, representing call center activity for PSTN migration campaign. It counts the significant responses from script questions, creates trouble tickets in the trouble ticketing system, and (un)successful migrations.

- IMS fault clearance daily report

    This report gives daily realization of fault clearance activity. It counts open/closed trouble tickets grouped by switch information, fault type, and service. It gives the efficiency of closed faults with or without taking the nonwork days into account. Efficiency in terms of line migration is monitored with columns that represent the line status (the line for which trouble ticket is raised): line status before a ticket is raised, after a ticket is closed, and the status when the report is generated.

- IMS fault clearance weekly report

  This report gives the weekly realization of fault clearance activity. It counts open/closed trouble tickets grouped by switch information, fault type, and service. It gives the efficiency of closed faults with or without taking the nonwork days into account. Efficiency in terms of line migration is monitored with columns that represent the line status (the line for which a trouble ticket is raised): the line status before a ticket is raised, after a ticket is closed, and the status when the report is generated.

- IMS fault clearance weekly summary report

  This report gives the weekly realization of fault clearance activity in pivot-like view. It counts open/closed trouble tickets grouped by switch information, fault type, and service. It gives the efficiency of closed faults with or without taking the nonwork days into account. Efficiency in terms of line migration is monitored with columns that represent the line status (the line for which a trouble ticket is raised): the line status before a ticket is raised, after a ticket is closed, and the status when the report is generated.

- IMS service provisioning daily report

  This report gives the daily realization of service provisioning activity. It counts open/closed SP work orders grouped by switch information and work order type. It gives the efficiency of realized work orders with or without taking the nonwork days into account. Efficiency in terms of line migration is monitored with columns that represent the line status (the line for which a work order is raised): the line status before a work order is raised, after a work order is realized, and the status when the report is generated.

- IMS service provisioning weekly report

  This report gives the weekly realization of service provisioning activity. It counts open/closed SP work orders grouped by switch information and work order type. It gives the efficiency of realized work orders with or without taking the nonwork days into account. Efficiency in terms of line migration is monitored with columns that represent the line status (the line for which a work order is raised): the line status before a work order is raised, after a work order is realized, and the status when the report is generated.

- IMS service provisioning weekly summary report

This report gives the weekly realization of service provisioning activity in pivot-like view. It counts open/closed SP work orders grouped by switch information and work order type. It gives the efficiency of realized work orders with or without taking the nonwork days into account. Efficiency in terms of line migration is monitored with columns that represent the line status (the line for which a work order is raised): the line status before a work order is raised, after a work order is realized, and the status when the report is generated.

- Lines status report (IMS)

This report gives detailed information for every new line that appeared in the system for a chosen date. It gives detailed information for the line (L mark, type, switch, geo location) as well as info for the last service provisioning work order and last trouble ticket order.

- Nonmigrated customers during fault clearance

This report gives information for trouble tickets of non-migrated customers. It is a weekly summarized report with information for nonmigration reason and switch.

- Nonmigrated customers during fault clearance detailed list

This report gives information for trouble tickets of non-migrated customers in a more detailed manner. It is a weekly summarized report with information for a nonmigration reason, switch, ticket type, line status, and number which verifies successful migration.

- PSTN migration trend

This is the main report for counting IMS lines. It takes weekly snapshots of the line situation in the network, and represents IMS and IMS-C (lines that are migrated to the IMS platform but still active on the PSTN platform) lines. It takes into account actions that caused the migration (if it is service provisioning, a trouble ticket, or proactive migration) and groups them accordingly. Additional information is of the work order/trouble ticket type.

KEY LEARNING

- Monitoring of processes and E2E migration for analysis.
- Identifying irregularities and creating corrective measures.
- Identifying key points and sensible parts of the process and detailed monitoring of them.

4.4 Interdependencies with Other General Projects

Identification of the interdependencies between this project and other projects on the IT/NT/corporate projects is crucial in order to identify the impact and update the project's scopes and deliverables.

4.4.1 Identification of the Interdependencies with the Current IT/NT Corporate Projects

- Review IT, NT, and corporate project roadmaps and identify the potentially affected projects.
- Review technical specifications and design of the solutions and identify the affected projects.
- Analyze implications of interdependencies.
- Adjust design of affected projects and align.

4.4.1.1 Identification of the Interdependencies with the Current IT/Corporate Projects Considering that the changes in the processes and systems for IMS implementation are significant, there is a high possibility for interdependencies and implications to other projects in the company. Therefore, all affected projects have to be identified and according to that, the technical specifications, budget plans, and alignment of the implementation timelines updated.

As a result, interdependencies are identified and can be managed by aligning the projects with PSTN migration. The following section provides an approach on how to identify and align those interdependencies with other projects.

4.4.1.1.1 Approach and Learning for Project Interdependency Identification and Management Considering all the changes that have to be implemented in the identified IT systems and interfaces between them in the different domains and the large scope of impact, the PSTN migration is likely affecting other IT and corporate projects. The impact could be in the scope, budget, and time of implementation and could directly influence the success of the project implementation. Thus, it is very important to analyze and identify affected projects and to ensure proper alignment of the implementation plans and budgets of the different projects.

There are several steps that have to be followed in order to identify the projects and to update the scope, implementation plan, and budgets in relevant cases:

1. Review IT and corporate project roadmap and identify the potentially affected projects.
2. Review technical specifications/design of the solutions and specify the affected projects.
3. Analyze implications of interdependencies.
4. Adjust design of affected projects and align the timelines.

The details for each of the steps are given in Figure 4.17, where all prerequisites, dependencies, and outputs are specified.

According to the companies' experiences, the steps have to be performed in the presented order as the outputs from one step are inputs for the other, risking limited NT/IT functionality, delays in project timelines, and cost increase:

1. In the first step, all current and planned projects should be collected. Project managers of the selected projects have to be aware about the PSTN migration project. They have to share this awareness with the vendors and system integrators of certain projects in order to actively participate in the process of identification of possible interdependencies.

Steps	Review IT and corporate project roadmap and identify the potentially affected projects	Review technical specifications and design of the solutions and identify the affected projects	Analyze implications of interdependencies	Adjust design of affected projects and align the timelines
Description	• Collect all current and planned IT/corporate projects • Check their timelines	• Review scope of the projects • Review technical design • Identify affected projects	• Identification of the technical dependences between the affected projects • Gap analyses • Risk evaluation	• Update of technical specification of affected projects and aligning the scope and timelines • Budget approval
Prerequisites	• All relevant project managers have to be aware about the IMS migration project • Availability of project timelines	• Technical design readiness • Project's plan readiness	• Readiness of detailed technical specification • Resource availability • System integrator/vendor's willingness to cooperate	• Gap analyses • Check if business case update is needed • Close cooperation between the different project teams
Dependencies	• Project timelines • Budget constraints	• Impact from IMS transformation project to the others	• Project's timelines • Project's scope	• Business requirements for project go-live • Budget availability
Result/ Milestone	• List of all current projects • List of future planned projects	• List of affected projects	• Gap analyses	• Updated technical specifications • Updated project plans • Updated budget plan

Figure 4.17 Approach and learning for project interdependency identification and management.

2. In the second phase, not only the scope should be reviewed, but also the timeline of the different projects, as the integration test on complete E2E processes needs to be performed with all changes by all projects implemented. All projects with potential impacts identified in this step need to be further analyzed and managed by the following steps.
3. The next step is identification of technical dependencies between the affected projects, gap analyses, and evaluation of risks in each project from the impacts for IMS implementation.
4. Gap analyses should be used for the final step, the adjustment of project plans and specification resulting in:

- Updated technical specifications for every affected project.
- Updated project plans aligned between the affected projects.
- Updated budget plan.

After these adjustments, the projects can continue with the implementation phases and finally with the successful implementation of the systems, requested functionalities, and efficient support of the PSTN migration processes.

According to country experience, the affected projects with the interdependencies from PSTN migration process implementation are the projects related to the:

- Systems from the affected domains listed in Section 4.3.1.1
- Interfaces that have to be modified from the affected domains listed in Section 4.3.1.2
- Affected processes are listed in Section 4.3.1.3

KEY LEARNING

- Consider all projects on the roadmap.
- Include IT architecture from the very beginning of the specification.
- Make the analyses before the financial planning cycle, and plan budgets accordingly.
- Create a big picture for all projects on the roadmap and align timelines regarding the interdependencies.
- Use a program management approach.

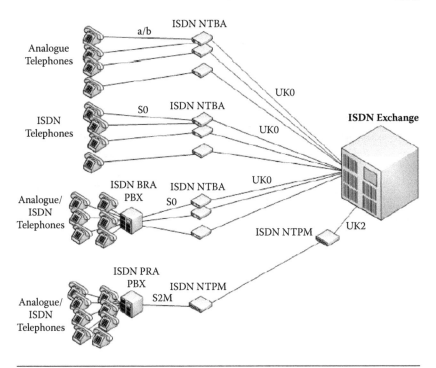

Figure 4.18 Legacy ISDN platform.

4.5 Evolution of the Network from Legacy (PSTN/ISDN) to the New IP Platform

As the old legacy system reaches end of service/end of life, economical and operational reasons enforce its replacement (Figure 4.18).

Three solutions for migrating the old platform to IP have been identified thus far.

1. CPE GW migration path (rollout of smart CPEs):

- The IP network ends at the customer's premise.
- A new HGW has to be installed at the customer.
- Customer has to be informed when the migration is carried out.
- Customer can use his old devices (Figure 4.19).

2. CO migration path (rollout of MSAN/ISDN cards) (Figure 4.20):

- The IP network ends in the exchange or in the box near the customer's premises.
- Customer does not always have to be informed when the migration is carried out.

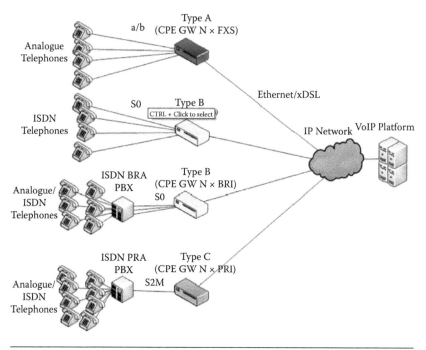

Figure 4.19 Smart CPE approach.

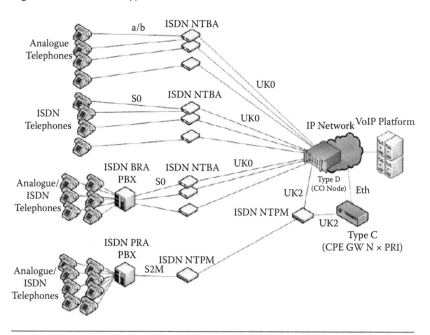

Figure 4.20 MSAN/ISDN card approach.

- Customer can use his old devices.
- Enables a cheap and fast migration.
- Low risk of churn.

3. All-IP migration path (Voice over Broadband, VoBB) (Figure 4.21):

- The whole network is IP based (all-IP solution).
- IP router has to be installed in each premise of the customer.
- In some cases new devices are needed for the customer.
- Future-proof solution.

The above-described migration paths do not represent stand-alone solutions. All options may be used in parallel within the migration plan, depending on the chosen migration strategy (Figure 4.22).

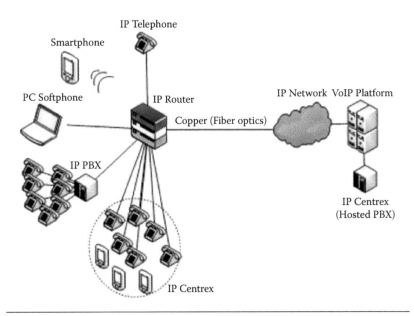

Figure 4.21   Voice-over-broadband approach.

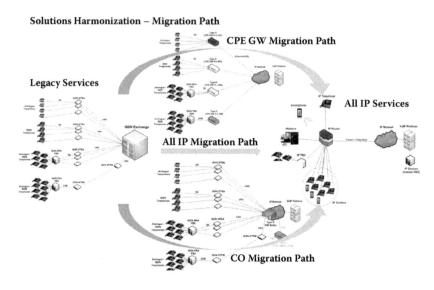

Figure 4.22 Integrated migration approach.

5

BUSINESS CASE FRAMEWORK

In all countries, PSTN migration is expected to have a sustainable impact on long-term platform cost development and to require significant investments for migration of the full customer base from the old PSTN legacy network to the IP world. The benefit of performing the migration needs to be calculated and monitored through a business case making a financial impact, as well as ambition transparent, to all involved functions in the project, local decision bodies, and other stakeholders.

The business case could simply be constructed by comparing migration costs to cost savings due to the shutdown of PSTN platforms. Experience shows that migration from PSTN to IP technology costs around 30€ ($38) to 60€ ($75) per subscriber, distributed over 3 to 5 years while the expected cost benefits from switching off the PSTN platform are around 10€ ($13) per subscriber per year. However, these figures only provide a rough orientation and one must bear in mind that

- It is often unclear which cost categories are migration related and thus in scope.
- Migration costs are heavily influenced by required resources in each customer segment, especially toward the end of cleansing an area.
- Business case results are highly sensitive to migration-related churn; already a small percentage of the total customer base churn can alienate the financial benefit of the business case.

Additionally, for decision-relevant business cases, the question—what to compare PSTN migration with—is not an easy one. Comparing it to an imaginary "flat line" do-nothing scenario is not realistic, as do nothing yields a strong rise in costs, as well as a high risk of network failures and a total failure of the PSTN in the long run. Thus, it is not a real option and the comparison of the PSTN migration scenario should be conducted against a realistic scenario or minimal change—usually a midterm interim solution such as prolonged PSTN usage or soft switches.

Hence, the business case has to be extended by taking various additional factors into account.

To support countries facing these challenges for a benefit calculation of investment projects, a harmonized business case framework has been developed as part of the PSTN migration blueprint with the following objectives:

- Provide a solid and comparable basis to orchestrate CAPEX, OPEX, and revenue discussion within and across all functions (e.g., marketing, sales, customer care, NT, IT, regulatory).
- Ensure cross-functional alignment for target setting during the conception phase of the project and rollout preparation as well as final execution of the migration plan.

The business case consolidates financial input from all work streams of the PSTN migration project (see Figure 5.1), aligned with national and group requirements for a business case calculation. Financial results of a business case calculation are relevant for decision support and commitment consolidation on the country level, but at the same time, figures will be used to create transparency at the group level (see Figure 5.1).

Input

PSTN migration project
- Overarching decision on scenarios, project timeline, and milestones
- Financial data from commercial roadmap (e.g., churn assumption, migration costs, cross- and upselling)
- Financial data from product portfolio roadmap (e.g., regulatory demands, product development costs)
- Financial data from NT/IT roadmap (e.g., IMS investment, technical product development)

Other input
- NatCo specific accounting rules
- Group controlling guidance on BC calculation

Figure 5.1   Business case—general input.

It needs to be emphasized that our harmonized business framework should be used for a benefit calculation of different scenarios in order to support "go/no-go" decisions on CAPEX investment for access technology and monitoring during realization on the level of each country.

A like-for-like comparison of business case results across different NatCos is hardly feasible due to different starting points depending on the status of the legacy network and specific migration scenarios. In particular, the availability and penetration of broadband ports may determine the operator-specific migration scenario. Other influential and country-specific factors include the status of the legacy network equipment for providing voice services (e.g., AXE, EWSD) and a forecast of equipment lifetime and maintenance costs.

Our business case framework supports NatCos by providing guidance on key questions occurring during a PSTN migration project. These questions will be addressed in the following sections:

- The definition of common calculation methods (e.g., business case runtime).
- Guidance on the calculation of business case scenarios (e.g., baseline, business case structure, level of detail).

5.1 Calculation Methods

The first challenge for setting up the business case for PSTN migration in a country is to clarify and define the parameters and scope for calculation.

This activity requires in detail a definition of basic calculation rules, scoping of cost items, treatment of parallel projects with (potential) overlaps, and rules for benefit calculation.

5.1.1 Calculation Rules

The determination of calculation rules includes all activities related to the definition of business starting point and runtime, calculation parameters, treatment of costs such as CAPEX or OPEX, and depreciations.

The starting point for the calculation should always be the first planned incremental investment into PSTN migration. Runtime for the calculation is defined as a fixed period of 10 years. This timeline should be defined in close collaboration with the person who is responsible for the overarching project as it determines the reference for calculating all possible scenarios, meaning investment into the migration or an alternative (interim) solution.

In general, the treatment of migration costs as CAPEX or OPEX, depreciation times for new equipment, and consideration of potential write-offs on legacy equipment have to be defined separately for each country to fulfill local accounting rules.

5.1.2 Scope of Business Case Items

The challenge is to get transparency on the benefit of PSTN migration for management decisions and monitoring during the execution phase. PSTN migration is necessary in order to ensure voice service revenues in a long-term perspective and is often related or runs parallel to other investment projects that are in rollout of fixed broadband (e.g., introducing xDSL broadband, FTTX).

It is recommended to include only incremental revenue and cost effects related to the migration itself in the business case in order to make the financial benefit of PSTN migration visible (Figure 5.2).

5.1.3 Financial Impact Calculations

The benefit of PSTN migration is determined by several factors that need to be taken into consideration from the start of the project. In order

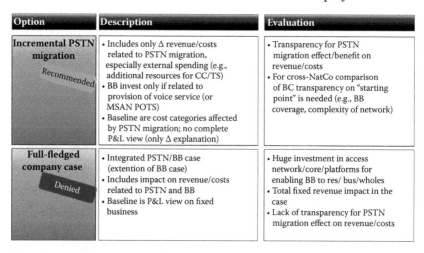

Option	Description	Evaluation
Incremental PSTN migration *Recommended*	• Includes only Δ revenue/costs related to PSTN migration, especially external spending (e.g., additional resources for CC/TS) • BB invest only if related to provision of voice service (or MSAN POTS) • Baseline are cost categories affected by PSTN migration; no complete P&L view (only Δ explanation)	• Transparency for PSTN migration effect/benefit on revenue/costs • For cross-NatCo comparison of BC transparency on "starting point" is needed (e.g., BB coverage, complexity of network)
Full-fledged company case *Denied*	• Integrated PSTN/BB case (extention of BB case) • Includes impact on revenue/costs related to PSTN and BB • Baseline is P&L view on fixed business	• Huge investment in access network/core/platforms for enabling BB to res/ bus/wholes • Total fixed revenue impact in the case • Lack of transparency for PSTN migration effect on revenue/costs

Figure 5.2 Scope of business case items.

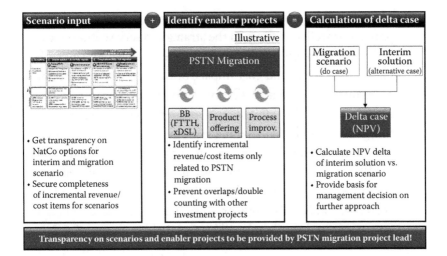

Figure 5.3   Steps to prepare benefit calculation.

to contribute to a straight investment and rollout decision, financial benefits and incremental impact on revenue, OPEX and CAPEX can only be calculated if the migration scenario and alternative options are well defined and if enabler projects are transparent (see Figure 5.3).

The person who is responsible for the business case has to consider the following in order to conduct a solid benefit analysis and to challenge the financial aspirations of the PSTN migration project:

- *Scenario input*: Each operator faces the challenge of defining a migration scenario and alternative interim solutions in an early stage of the project. Depending on the status of the legacy network and broadband, scenarios have different implications on revenue, and OPEX and CAPEX assumptions that need to be included in the business case. To give an example, migration of voice-only customers via MSAN POTS port avoids churn risk for this specific customer group but inhibits the chance for cross- and upsell during the migration process at the same time. Further details on this scenario definition will be explained in Section 5.2.1.
- *Enabler projects*: Key prerequisites for business case calculation are the total transparency on the execution of enabler projects or setting up new strategy directions regarding voice and broadband service in a given market. For example, execution

of predicted broadband penetration is one of the key enabling projects as well as setting up the strategy to serve certain locations only through mobile services. The financial effects with regard to revenue and cost impact of those projects must not be counted again in a PSTN migration business case.

- *Delta case*: An investment and rollout decision for PSTN migration should finally be made through delta NPV calculation of incremental revenue, operational cost development, and CAPEX spending for interim solution versus migration scenario. The respective input has to be provided by commercial, product portfolio, and NT/IT work streams. The required level of detail for revenue and cost items will be explained in Section 5.2.2.

However, it is important to note that the delta case is calculated to make a decision on PSTN migration or an alternative interim solution and to show the differences in financial impact. But a delta case must not be used to show a financial impact on absolute revenue or the cost base of a concrete financial year (e.g., comparing revenue and cost position after migration to the year before the project start, or any other year during the migration), because the delta figures cannot be shown or found in a P&L view. Thus, for reporting purposes only the financial impact of the chosen scenario (usually migration scenario) can be used (see Figures 5.4 and 5.5).

KEY LEARNING

- Transparency of related investments in a fixed network infrastructure are fundamental for the calculation of a PSTN migration benefit.

5.2 Project Assumptions

This section provides guidance on structure and the level of data needed for the calculation and information on business case assumptions from all work streams as defined in the PSTN migration blueprint.

It has to be mentioned that the calculation of the migration benefit is as of today primarily about cost avoidance and churn minimization. Hence, the challenge is especially about the identification of the

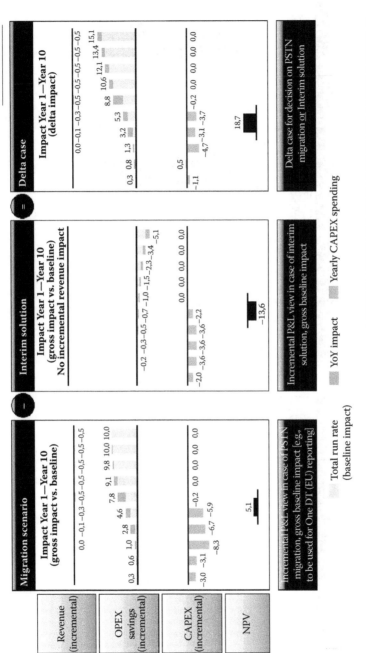

Figure 5.4 Delta case calculation and P&L view per scenario.

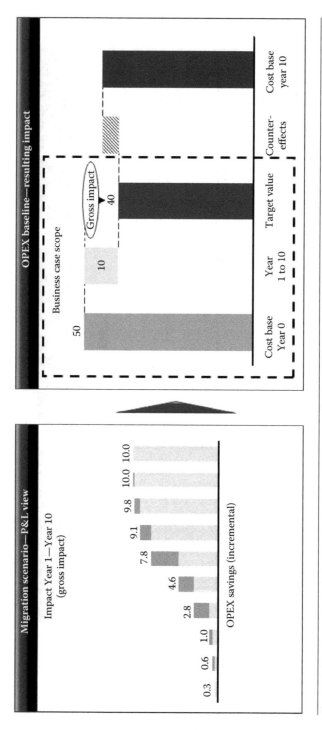

Figure 5.5 Example of a migration scenario—gross OPEX impact on cost base in P&L.

major drivers of migration costs, operational cost development, and revenue loss from migration-related churn or regulatory moves.

Although PSTN migration is often associated with ideas for new IP services, the long-term impact on revenue is currently difficult to foresee. At any rate, for consideration of those revenue effects, the business case framework provides the possibility of including incremental revenue effects from new services or cross- and upselling during the migration process.

For this activity, the business work stream needs transparency on the overarching project timeline, the NatCo's specific scenarios, and resulting input needs from work streams:

- *Scenario definition*: Operator's specific interim solution (i.e., prolonged PSTN usage or soft switch investment) and migration scenario (i.e., MSAN POTS for voice only or broadband for all customers).
- *Project timeline and milestones*: Milestones for business case calculation have to be aligned with overarching project timeline and management board decision points.
- *Input migration*: Financial data with regard to revenue, OPEX and CAPEX impact for NatCo's PSTN migration scenario (including timeframe and drivers/levers), for example, churn assumptions, migration costs, and long-term impact on fault clearance.
- *Input product portfolio roadmap*: Financial data with regard to revenue, OPEX and CAPEX impact for NatCo's PSTN migration scenario, (including timeframe and drivers/levers), for example, product development costs, revenue up- and downsides.
- *Input NT/IT roadmap*: Financial data with regard to revenue, OPEX and CAPEX impact for NatCo's PSTN migration scenario (including timeframe and drivers/levers), for example, technical product development costs, access and core network investment as well as energy costs, and maintenance and personnel expenses.

5.2.1 Scenario Definition

As described above, the "go/no-go" decision for PSTN migration has to be made by comparison of revenue and cost development of

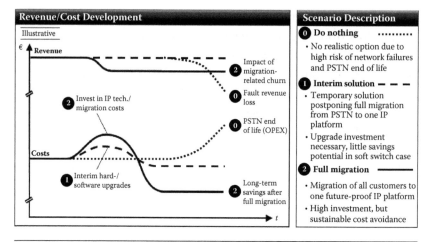

Figure 5.6   Illustrative revenue/cost development for scenarios.

different country-specific scenarios. The definition of the scenario[*] itself is not the task of the business case work stream team, but it needs to have a good understanding of their revenue and cost implications to be able to calculate and challenge benefits of the migration (Figure 5.6).

In order to define the optimal scenario for further evolution of voice service and/or securing existing revenue from voice service, the following options are possible:

- *Do nothing*: Keeping the existing legacy network with nothing other than maintenance investments may result in a strong rise in costs, as well as in a high risk of network failures and a total failure of the PSTN in the long run.

- *Interim solution*: This option is defined as the minimum requirement to prolong PSTN usage with necessary platform upgrades or implementation of a soft switch. CAPEX is mainly needed for hardware/software upgrades or an introduction of a soft switch. Changes in OPEX may be related to *service level agreements* (SLAs), energy costs, and personnel expenses, but need to be checked carefully.

- *Full migration*: This scenario considers migration of all voice customers to all IP platforms in an area, either by migration

---

[*] For details and responsibility see overarching project blueprint/scenario analysis in Section 1.4 in Chapter 1.

of all customers via broadband or via MSAN POTS for voice-only customers. From the perspective of future proof of the solution, full migration is the optimal one, since it provides possibilities for convergent IP services (fixed and mobile) and even given the fact that at this moment it is difficult to identify new revenues only based on the voice services, implemented technology gives the possibility of developing options for new revenue streams as bundles with video and data services. Makedonski Telekom used this approach to migrate all customers to the IP services, including single voice customers. This scenario can be adapted by a combined solution, meaning not all areas are migrated to IMS within the given timeline (e.g., some PSTN switches have longer life time or—depending on operator strategy—some areas to be connected via mobile solution instead of fixed lines).

As described above, the benefit of PSTN migration versus an alternative scenario needs to be calculated as a delta case by comparing incremental financial implications of a country's specific interim and full migration options with regard to revenue, operational cost development, and CAPEX investment (see Figure 5.7).

Having that general understanding of scenarios to be calculated, the business case framework helps to align the financial impact from all other elements by area, customer segment, and product during project runtime (see Figure 5.8).

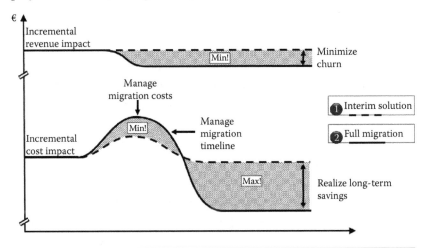

Figure 5.7 Business case delta and main drivers.

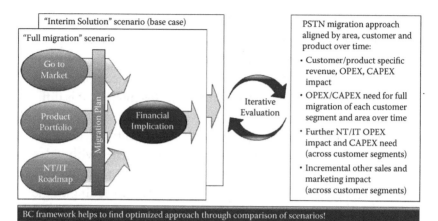

Figure 5.8   Business case framework—link to other elements.

5.2.2 Input Need from Work Streams

5.2.2.1 Milestones of Business Case Calculations The business case work stream has to set milestones during the PSTN migration process to increase maturity of calculation and to enable monitoring (see Figure 5.9).

Following a defined PSTN migration process (as described in the overarching topics section, see Section 1.1 in Chapter 1), three key maturity milestones and gateways have been defined accordingly:

- *Rough estimation*: This milestone is needed for getting a first insight on the financial impact of PSTN migration. Only mandatory major drivers are required but inputs and assumptions must be valid. However, the given business case structure should be filled with all available data at that time and reflect the commitment of respective functions or work stream.
- *Rollout decision*: Second milestone requires complete transparency regarding financial impacts of the PSTN migration based on high maturity of all inputs and drivers (bottom-up validation). This business case will be used to get the final commitment of all functions and the management board for the rollout decision.
- *Monitoring and annual financial planning review*: Once the execution of PSTN migration has started, regular monitoring of the business case is required by comparison of actual and planned values. This business case version should be updated in general once per year or in line with some special event or

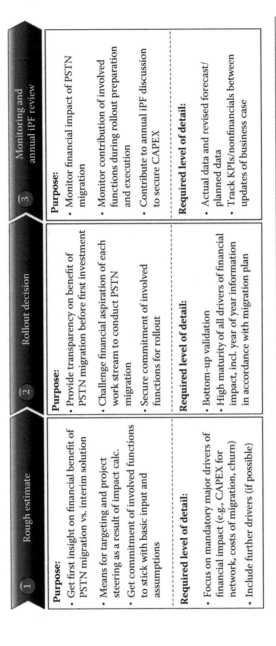

1 Rough estimate

Purpose:
- Get first insight on financial benefit of PSTN migration vs. interim solution
- Means for targeting and project steering as a result of impact calc.
- Get commitment of involved functions to stick with basic input and assumptions

Required level of detail:
- Focus on mandatory major drivers of financial impact (e.g., CAPEX for network, costs of migration, churn)
- Include further drivers (if possible)

2 Rollout decision

Purpose:
- Provide transparency on benefit of PSTN migration before first investment
- Challenge financial aspiration of each work stream to conduct PSTN migration
- Secure commitment of involved functions for rollout

Required level of detail:
- Bottom-up validation
- High maturity of all drivers of financial impact, incl. year of year information in accordance with migration plan

3 Monitoring and annual iPF review

Purpose:
- Monitor financial impact of PSTN migration
- Monitor contribution of involved functions during rollout preparation and execution
- Contribute to annual iPF discussion to secure CAPEX

Required level of detail:
- Actual data and revised forecast/planned data
- Track KPIs/nonfinancials between updates of business case

Figure 5.9 Project assumptions for business case framework.

	Interim Solution			Migration Scenario (Full or combined approach)		
	1 Rough estimate	2 Rollout decision	3 Monitoring	1 Rough estimate	2 Rollout decision	3 Monitoring
Migr. Plan & G2M						
•Migration plan (incl. churn, effort, investment per cust.)	✗	✗	✗	✓	✓	✓
•Communication plan	✗	✗	✗	✓	✓	✓
•Sales channel strat.	✗	✗	✗	✓	✓	✓
Product Portfolio Rm.						
•Revenue impact	✗	✗	✗	✓	✓	✓
•Product develop.	✗	✗	✗	✓	✓	✓
NT/IT Roadmap						
•Techn. product development	✗/✓ [1]	✗/✓ [1]	No more updates of interim impact calculation after IP rollout decision	✓	✓	✓
•NT/IT investment	✓ [2]	✓ [2]		✓	✓	✓
•OPEX impact (energy, SLA, FTE)	✓ [2]	✓ [2]		✓	✓	✓

Figure 5.10   Input need for business case—overview.

request and contribute to the annual financial planning discussion at the NatCo and Deutsche Telekom AG Group level.

This means that all work streams have to make their assumptions and dedicated levers transparent. Input from the migration plan and product portfolio roadmap is limited to incremental effects on the migration scenario as an interim solution usually does not require migration at customer premises or new product development. On the other hand, the NT/IT roadmap has to provide data for both scenarios: The interim solution requires assumptions on CAPEX need for prolonged PSTN usage (e.g., software updates) or investment into new soft switches and long-term operational cost development for maintenance, energy, and personnel expenses (see Figure 5.10).

5.2.2.2 Input for the Migration Scenario

5.2.2.2.1 The Migration Plan   Inputs from the migration plan are crucial for developing a transparent and comprehensive business case. Even in the rough estimate stage it should be clear how each of the inputs are calculated and how they influence the current company setup. The following drivers from the migration plan are considered major ones:

• Churn (run-rate revenue, direct cost effect) can be a major negative factor in PSTN migration. It should be carefully considered but not mixed with "natural" customer churn over years, for example, due to increased competition on the market. Churn from PSTN migration can come from recontracting (if substitute products need a new contract with the customer, or if general terms and conditions do not cover VoIP), from force migration process (customer unwillingness to migrate), or from bad quality of service perspective after migration (PSTN versus IP). This driver needs to be calculated per customer segment and by products.

- The NT workforce costs per migrated customer (OPEX or CAPEX depending on accounting policy) represents costs from needed work on customer premises including traveling costs, physical work, and wiring (material) cost at customer premises. It does not include customer premise equipment (CPE). This driver needs to be calculated per customer segment and aligned with the migration plan.

- The NT impact from a fault clearance perspective cost (run-rate OPEX increase or decrease) is based on the prediction that migrated voice-only customers will have shorter MTBF in comparison with PSTN due to the current data from MTBF in the broadband world and that this will be manifested as a higher number of faults compared with PSTN. The prediction is that by adding broadband port and CPE for a voice-only customer in one migration scenario or adding MSAN POTS port for a voice-only customer in another scenario will influence the current MTBF for voice customers. For prediction purposes one can use current experience from the structure of MTBF from broadband customers (what percent of faults happens due to broadband port and CPE). Special focus on this topic should be if the operator is using the migration scenario over broadband port and CPE for voice customers.

- Zero touch provisioning (run-rate OPEX savings) in the case of a scenario when the operator is using broadband port and CPE for migrating voice-only customers. By using this scenario, upgrading the customer on broadband service will be done remotely without technicians and the customer will be provisioned almost instantly. The same analytic can be used for OPEX savings in cases when a 2P customer is churning with broadband service. Zero touch provisioning gains should be in line with financial planning figures (gross adds and churn prediction) and in line with the PSTN annual migration plan.

Special attention should be given to transparency on the migration plan per customer segment. Migration types are divided as follows.

- Automigration represents one of the most efficient ways of migration. There are several ways of the automigration process:
  - By letter means that the customer receives the letter with all necessary details of the migration process and a toll-free 1-800 number. The customer can call customer care and do the migration following the steps described in the letter.
  - By IVR means that the customer has received the letter; customer care is trying through IVR to "remind" the customer to perform migration following the steps described in the letter.
  - By call center means that agents are trying to get in contact with the customer and perform migration with agent guidance.
  - Web based can be divided into two separate activities. The first one is that customers use an official operator portal to perform migration. The second one uses the "pop-up" principle (wall garden) where the customer is reminded to do migration each time that he tries to connect to the Internet.
- Supported migration is done with the company's own technicians or subcontractors. Basically, technicians or subcontractors must visit the customer in his premises and perform migration. This is the most expensive migration type.
- Two-in-one migration represents customer migration by technicians or subcontractors but together with provisioning or a fault repair process on the same customer. In this type, by one visit to the customer, technicians or subcontractors are performing two actions without additional incremental costs for migration.
- Migration via MSAN POTS port represents the fastest and cheapest migration of voice-only customers. On the other hand, by this migration type one is not entering any additional value into the network (just replacing the PSTN port for MSAN POTS port).

In addition, it is essential to critically check and review the input from all streams as the migration plan integrates the cross-functional aspects throughout the NatCos.

5.2.2.2.2 Product Portfolio Roadmap   Financial inputs from the product portfolio roadmap cover revenue gains from new features/services and revenue loss in case of certain regulatory decisions. Major inputs needed:

- Revenue downsize from existing services (run-rate revenue, direct cost effect) is based on assumptions with a high probability of occurrence and that PSTN migration is causing it. This mainly covers expected regulatory decisions (e.g., interconnection fee influencing revenue and direct cost). However, it should be clear that migration is not used as an excuse for regulatory decisions that would happen regardless of PSTN migration, hence not causing an incremental effect.
- Revenue potential from new features/services (run-rate revenue, direct cost effect) must be transparently calculated and represents additional benefit from PSTN migration. This driver needs to be calculated per customer segment and by products.
- Revenue loss from benefits (one-time revenue effect) needs to be calculated if benefits are given to customers in order to push or increase the likelihood of successful migration to IP (e.g., free minutes). This driver needs to be calculated per customer segment.

In addition, the product portfolio work stream has to be challenged with regard to product development costs related to the migration, either for existing products (1:1 substitution of features) or new services, and retirement costs of services (e.g., regulatory-required information to customers).

5.2.2.2.3 NT/IT Roadmap   Inputs from the NT/IT roadmap are crucial for completing the transparent investment picture for the full migration scenario, especially as most of the savings from operational costs are expected from this stream. The following drivers are considered as major ones:

- Access/core network investment (CAPEX) represents additional costs in the areas where there is no existing MSAN node (no broadband service available) and one is not planned to be deployed as part of the operator's overall fixed broadband business case. For such locations, all investment costs for deploying MSAN must be calculated as incremental in the PSTN migration business case because deployment of the MSAN will be done just for PSTN migration purposes.

- Core network platform investment (CAPEX) represents all investments in necessary platforms (IMS and business application servers) and other core network elements in order to support PSTN migration for each customer segment.
- IT process system investment cost (CAPEX) represents the cost of introducing new IT systems only in order to support PSTN migration.
- CPEs by customer segments (CAPEX) represent investment for enabling migration of voice-only customers (if such a scenario applies), for migration of existing broadband customers with inadequate CPE (i.e., current CPE without VoIP functions), and for future fault repairs in cases when voice-only customers are migrated over CPE and broadband ports. Additionally, price erosion over the years should be considered.
- MSAN POTS port for voice only (CAPEX) represents investment in the central office in order to support migration of voice-only customers via MSAN POTS ports (if such a scenario is applied). Additionally, price erosion over the years should be considered. This driver needs to be calculated per customer segment.
- xDSL port costs (CAPEX) represent investment in the central office in order to support migration of voice-only customers through CPE and broadband ports (if such a scenario is applied). Additionally, price erosion over the years should be considered. This driver needs to be calculated per customer segment and it should be checked carefully if the required CAPEX is within the scope of the broadband case.
- IMS license (CAPEX) investment costs are directly linked with the number of customers that need to be migrated. License costs should be in line with financial planning figures for the annual total number of migrated voice customers. This driver needs to be calculated per customer segment.

- NT SLA savings (run-rate OPEX impact) represent the difference between current NT SLA cost for PSTN (local exchanges, RSU) and NT SLA after migration is done (IMS, MSAN). NT SLA costs after migration must consist of just cost that occurs due to the PSTN migration (i.e., IMS, part of MSAN SLA referring to additional ports needed for migration). NT SLA cost development needs to be linked closely with the annual migration plan.

- Energy savings (run-rate OPEX impact) represent the difference between energy consumption for PSTN (e.g., local exchanges, RSU, cooling) and energy consumption after migration is done (e.g., IMS, MSAN, cooling). The baseline for PSTN energy consumption must be carefully defined due to the fact that today, in most cases, there is no clear line in current central office architecture between PSTN and other technology (i.e., xDSL equipment, transmission equipment, mobile equipment, IT equipment) consumption, especially regarding cooling. Regarding consumption after PSTN migration one must take into consideration not just the increase of energy consumption that occurs due to the PSTN migration project (IMS and other platforms, additional MSAN ports [POTS cards and broadband ports], additional cooling). Energy cost development needs to be linked closely with the annual migration plan.

- NT personnel expenses (run-rate OPEX impact) represent savings due to the fact that after PSTN migration there will be no further need for a certain number of FTEs (or external workforce) in the domain of planning and assembling MSAN technology as well as in maintenance of PSTN. OPEX gains from this aspect of PSTN migration could be one of the major inputs for complete business cases. One should not forget to calculate the necessary number of FTEs needed for IMS to have an incremental efficiency impact.

- Rental savings (run-rate OPEX impact) represent additional benefits from PSTN migration due to the large space (sqm) that will be freed up after dismantling old local exchanges and RSU. From the beginning of the migration process one should perform a deep analysis of the existing premises where PSTN migration takes place. It is necessary to have a clear picture of rented premises. These values must be shared with the technical planning department in order to secure that the technical solutions for each central office include a future real estate function.

- Revenue from selling or renting premises (run-rate or one-time revenue) represents additional benefits from PSTN migration due to the large space (sqm) that will be freed up after dismantling old local exchanges and RSU. From the beginning of the migration process one should perform a deep analysis of the existing premises. It is necessary to have a clear picture of the owned premises and their value in the market. These values must be shared with the technical planning department in order to secure that the technical solutions for each central office include a future real estate function.

- The NT overarching workforce costs for field service (one-time OPEX or CAPEX depending on the accounting policy) represent costs from assembling the MSAN equipment in central offices including assembling of hardware, wiring costs, cooling and power hardware costs, and physical work costs (including travel expenses to the central office) of the company's own technicians or external partners. MSAN hardware costs are excluded.

- Special depreciation of inventory represents write-offs of dismantled equipment (local exchanges, RSU, and inadequate CPE) during PSTN migration. Write-offs must be aligned with the annual PSTN migration plan and local accounting policy regarding EBITDA (earnings before interest, taxes, depreciation, and amortization) impact.

In addition, the NT/IT work stream has to be challenged, for example, with regard to product development costs related to the migration, either for existing products or new services or IT SLA costs.

5.2.2.2.4 Go to Market Inputs from go to market are crucial for the full migration scenario, especially as here the customer base is touched. The following drivers are considered as major ones:

- Up- and cross-selling (run-rate revenue, direct cost effect) potential during PSTN migration represents a major opportunity for revenue and direct cost increase. Customers might be contacted several times through several sales channels (technicians, customer care, door-to-door agents, and indirect partners) and should be exploited. In the case where operators choose a scenario of migrating voice-only customers over broadband port and home gateways, upselling from 1P to 2P should be a major focus. This driver needs to be calculated per customer segment and by products.

- Customer care workforce costs (one-time OPEX[*]) should be evaluated from several angles. The right combination of outgoing calls from agents versus IVR should be used. Furthermore, the number of calls per customer also creates additional costs that should be considered as well as the number of incoming calls for migration. However, customer migration through customer care represents one of the most cost-efficient models. This driver needs to be calculated per customer segment and aligned with the migration plan.

- Customer care long-term perspective costs (run-rate OPEX increase or decrease) are based on the prediction that migrated voice-only customers will have a shorter mean time between failure (MTBF) in comparison with PSTN due to the current data from MTBF in the broadband world and that this will be manifested as a higher number of calls to customer care compared to PSTN. Basically, the prediction is that by adding a broadband port and CPE for a voice-only customer in one migration scenario or adding an MSAN POTS port for a voice-only customer in another scenario will have an influence on current MTBF for voice customers. For prediction purposes, one can use current experience from the structure of MTBF from broadband customers (what percent of faults happens due to broadband ports and CPE). Special focus on this topic should be considered if the operator is using a migration scenario over a broadband port and CPE for voice customers.

Furthermore, market communication costs for existing and new products, expenditures for postal charges, sales costs, and training

---

[*] Only noncapitalized costs.

costs should be checked, although they are not in general considered to be major drivers of financial benefits.

5.2.2.3 Input for Interim Solutions All input for the calculation of an interim solution needs to be provided by the NT/IT work stream, by collecting the following inputs.

- Network update (upgrade) costs (CAPEX) represent all investments in soft switches or in upgrading existing local exchanges with new software versions.
- Migration costs (one-time OPEX) for installation of soft switches (part of installation costs that will be treated as OPEX, e.g., traveling costs).
- Installation costs (one-time OPEX or CAPEX depending on local accounting policy) in case of soft switch scenarios focusing on installation of MSAN H248 in the central office.
- NT/IT product development (one-time OPEX) in the case of soft switches if technical adaptation is required to provide the product portfolio to customers.
- NT SLA costs (run-rate OPEX impact) represent the difference between current NT SLA costs for PSTN (local exchanges, RSU) and NT SLA after the interim solution is implemented (soft switch or local exchange upgrades). NT SLA cost development should be linked with the annual migration plan.
- Energy savings (run-rate OPEX impact) represent the difference between energy consumption for PSTN (local exchanges, RSU, cooling) and energy consumption after the interim solution is implemented. The baseline for PSTN energy consumption must be carefully defined due to the fact that today, in most cases, there is no clear line in current central office architecture between PSTN and other technology (xDSL equipment, transmission equipment, mobile equipment, IT equipment) consumption especially regarding cooling. Energy cost development should be linked with the annual migration plan.
- Rental savings (run-rate OPEX impact) represent additional benefits from an interim solution (soft switch scenarios) due to space that will be freed up after dismantling old local exchanges/RSU.

KEY LEARNING

- Align the structure for input data in the early stage of the project.
- Transparency on calculation assumptions is needed.
- Take accounting rules into consideration (e.g., treatment of CPE installation, free IMS support versus fair value calculation).
- Keep an eye on changes of overarching migration timeframe and interdependencies.

5.3 Business Case Structure

Finally, this chapter gives guidance on business case structure for benefit calculations from interim versus full migration scenario delta case.

For this purpose, all parameters and inputs defined in the previous chapters on common calculation methods and project assumptions need to be put into one structured business case Excel file for benefit calculations. Therefore, it is required to define the absolute cost and revenue base before project start (equaling PSTN related actuals in the year before PSTN migration starts) as well as baseline items for calculation, setup business case structure, and Excel files.

- *Define absolute revenue and cost base items*: Definition of base revenue and cost items that have to be transparent as a starting point for impact calculation of both migration and interim scenario.
- *Business case structure*: The input from all work streams needs to be put into structured P&L business case view, allowing analysis by customer segments and functional categories. This structure is 1:1 reflected in an Excel tool.

KEY LEARNING

- Collect backup information for all business case input to secure transparency on assumptions (e.g., volume per migration type and assumed costs per migration type).
- Get sign-off from all involved functions/work streams to achieve commitment for benefit realization.

5.3.1 Revenue and Cost Base and Business Case Baseline

Revenue and cost base: Starting position has to be defined in absolute terms for all relevant revenue/direct/indirect cost categories with incremental run-rate effect either from PSTN migration or an alternative interim solution. It depends on the starting point of the calculation, that is, the base has to be set 1 year before the first incremental investment in migration scenario happens.

Business case baseline: The business case baseline, that is, the cost and revenue development over time in case of "do nothing" (maintaining PSTN platform as long as possible), needs to be developed in absolute terms and on a per-item level (e.g., per SLA type), reflecting among others assumed annual price increases, inflation rates, churn, changes in quantity consumptions, et cetera. This baseline needs to, among others, calculate the incremental deltas for both BC scenarios—interim solution and full migration scenario—that determine the scenario's NPV.

Due to the fact (see also Section 5.1.3, "Financial Impact Calculations") that the incremental business case view cannot be used to show absolute cost developments over time and calculate savings compared to historical actuals, it is necessary to document the annual absolute cost development for all scenarios (baseline, interim solution, full migration scenario). Comparing the annual absolute development per each quantified item (absolute full migration scenario versus absolute baseline scenario = BC full migration; absolute interim scenario versus absolute baseline scenario = BC interim scenario) delivers the increments needed for the NPV calculation.

6

Go to Market

This chapter provides an in-depth look at the commercial aspects of the PSTN migration process. The main activities of the go-to-market work stream include:

1. *Communication plan*: Delineation of the external and internal communication strategy and plan.
2. *Sales channel plan*: Orchestration of the sales channels in the migration process and up- and cross-selling opportunity.
3. *Training*: Development of training activities for all functional groups that are involved, including an all-employee training.
4. *Migration support from CC/technical service*: Optimization of resources needed during and after the migration.

6.1 Communication Plan

This section provides guidance on what aspects to take into account to set up communication for PSTN migration with regard to the definition of a communication strategy and communication plan.

6.1.1 Communication Strategy (Including Benefit Story)

To present the PSTN migration in a positive way to the public and internal stakeholders, a communication strategy should be built. With regard to the customer, this usually means incorporating a benefit story into the communication strategy, which encompasses superior products with a unique selling proposition (USP) or a strong promotional offer/proposition.

The existing PSTN network is reaching its end of service, and it has to be replaced with a new technology to ensure future services. Setting up a communication strategy and a benefit story is complicated by the fact that, as of today, IP technology has no clear and communicable benefits over the existing legacy platform, at least

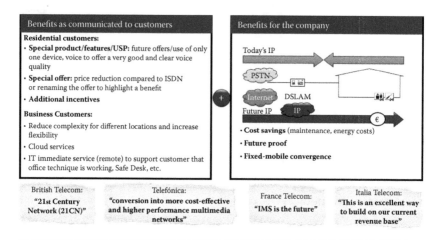

Figure 6.1   Benefits of PSTN migration for the customer and company.

from a customer perspective. Hence, the main point of an external communication strategy should be that the company is responsible for its customers and this is shown with a large modernization by installing a future-proof network for a higher quality and service standard.

In order to convince customers and to increase willingness for migration, some additional benefits or incentives as part of IP product bundles are strongly recommended (see Figure 6.1). In cases where this convincing might lead to revenue loss, new products/bundles/offers should be prepared upfront: in this way, a new IP offer is sold to the customer rather than the migration itself, which will then be conducted at the same time. For the internal stakeholders, the migration from a PSTN to an IP network is beneficial not only for NT/IT functions, but for the company as a whole and thus much easier to communicate.

However, there are also barriers as current customer perceptions on VoIP and related services are rather negative. Combined with additional barriers from the company's perspective, the PSTN migration can turn out to be more troublesome and slower than expected.

Therefore, the benefit story needs to overcome hurdles in the migration process to IP. Table 6.1 provides examples from different countries that were used in communication to overcome customers' resistance to migration.

However, experiences from countries undergoing the migration have shown the need for additional stimulus. In the situation where there is no clear benefit for the customer from the introduction of

Table 6.1 Messages for Communication—NatCo Examples

NatCo	MESSAGES FOR COMMUNICATION
Macedonia	• High-quality convergent multimedia services
	• Innovative ways of usage of the unlimited possibilities offered by the Internet
	• Macedonia is the first country in Southeast Europe whose network will be fully based on Internet protocol
	• The modernization of the networks will further strengthen their capacity
	• This is a promise that we have made to the customers and we are already fulfilling that promise
	• This project will provide the citizens with telecommunications solutions of the next generation
	• All communication devices that can be used at home will be connected to this device—home telephone, MaxTV, computer, laptop, smart phone, as well as all new types of TVs, Web cams, Internet radios, etc.
	• In the future, it shall be possible to manage or control all such home devices through the Internet
	• It will enable access to the new communication services and solutions with superior quality
Croatia	• Benefits mentioned: Faster activation of additional services, shorter period needed for installation of broadband services (e.g., IP TV), remotely access in case of upgrading certain services
	• Migration to new technology is needed to ensure better services
	• IMS—the most advanced technology of fixed telephony
	• IMS is a future technology that enables the use of advanced services in fixed telephony and a simpler and faster way of providing a variety of innovative services
	• New technology, in the future, brings enrichment of offers based on the Internet
	• The migration to new technology is completely free and easy to do
Germany	• Connect-Test (issue 02/2011): "Good"—biggest telecommunication magazine in Europe
	• With the IP-based access you are up-to-date with telephony of the future
	• You will receive many features without additional costs, e.g., 2 voice lines and 3 numbers
	• The telephony center in the Internet will allow you to change your phone settings, e.g., activate a call diversion
	• Additional features such as rejection of unwanted callers are without additional costs
	• You can do your settings via the Internet by yourself
	• You can use your home line wherever you are

the new product, the focus is on other aspects, for example, linking the communication message to overarching corporate strategy and further more specific propositions to the customers.

One country linked its PSTN migration strategy with the overall corporate strategy. Being a market leader and innovator, the new platform was introduced for their customers to enjoy next-generation services, taking a step forward from the digital network. Therefore, they clearly highlighted advantages for the customer for the entire transformation process. All the benefits in the marketing message were explained to the customer from the technical side, following the overall strategy, showing customers that they were the ones who benefited from the migration. The emphasis was put on new, multiple connected services at home/office that would facilitate customers' lives and work, a simple installation, and no cost.

Other countries, for example, introduced more features (comparable to ISDN) for the same price such as the old legacy product—therefore a clear customer benefit could be communicated, even targeted with certain groups (e.g., families with babies could use two lines operating like a baby phone).

To summarize, a direct link to the overall strategy helps to ensure the consistency, reliability, and fulfillment of customers' expectations. PSTN migration, strongly built on the existing corporate strategy, should be combined—if possible with regulatory requirements—with special offers, for example, special price offers or new services for limited time, strengthening the benefit story.

To stimulate even higher success for the migration, additional benefit stories should be developed for each customer segment, targeting product and features, price, additional incentives, and target group-related offers.

Pure voice customers can also benefit from the launch of new IP products, which are available only with a full VoBB strategy. Along these lines, Makedonski Telekom developed Broadband on demand, which offers additional benefits to the customers, as higher flexibility and immediate activation, and increases the revenue per customer. Further details of Broadband on Demand are described in Section 6.2.2, Figure 6.6.

Finally, market research is a necessary tool for the company to identify customer satisfaction and to promptly react to the negative perception of users. These studies should be conducted in all the migration phases (natural, active, and forced) and equally monitored for customer satisfaction on products, transactions, and corporate loyalty. For a definition of the different phases, please check Chapter 2, Section 2.2.

6.1.2 Communication Plan

In the next step, the communication approach and timeline have to be planned. Customers should receive general and specific information on time, in order to push cost-efficient natural migration (NM) from an early stage of migration execution.

6.1.2.1 External Communication Communication directed at migrating customers begins before the official migration starts and is managed in a stepwise approach. First, during the rollout preparation and shortly after the official start, it plays an informative role and aims to create positive market awareness and provides general information to the customers, thus educating the market and different customer segments. Afterward, it serves as support for all migration phases with communication materials, becoming more and more targeted for each customer toward the stage of forced migration (FM) (Figure 6.2).

According to the migration strategy chosen, communication channels need to be orchestrated in different phases (natural, active, or forced) in order to define which channels are the most suitable for each of the phases and at which point in time the marketing material should be released. It should be taken into account that each customer group requires a different communication approach. The education of the market represents an important condition for the migration success. Generally, customers are unaware of technology changes and this lack and/or scarcity of communication intensifies the resistance and skepticism of consumers. Therefore, it is necessary to provide coherent and clear information from the early start of the

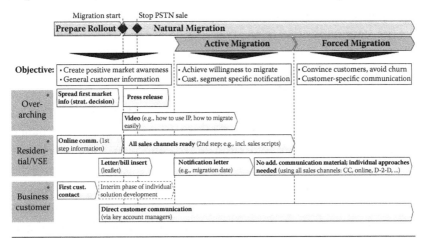

Figure 6.2 Communication roadmap—general approach.

migration. This education of the market should be carried out for both residential and business customers, using the preferred channels for spreading notifications.

KEY LEARNING

For a successful migration start and communication plan setup, the following prerequisites are necessary before the communication to the market starts:

1. Establishment and smooth execution of the processes.
2. Availability of substitute products on the IP network.
3. Successful pilot run prior to migration start.

6.1.2.1.1 Residential Customers   The communication material, defined for each customer group and phase, requires a specific timeline, which should include:

- Time required for preparation of each particular promotional material (production).
- Time required for possible testing and verification.
- Timeframe in which the material will be communicated to the customer (communication).

The start of PSTN migration is a focal point in planning a timeline for each communication channel.

The communication plan must consider the different migration phases (natural, active, and forced) and their communication objectives, such as information or call to action. Furthermore, each migration phase requires regional specification.

Usually, communication is focused in the first step on the natural migration phase, informing customers nationwide about the change of the platform via TV and radio shows, online articles, and letters or leaflets to the customer. Later, the countries actively approach the customers to call for action.

In the following section, we highlight examples of communication material that have been used in four countries during the natural and active migration phases.

NATURAL MIGRATION PHASE

When starting with the migration, the first step is to inform and educate the market. This communication, therefore, takes place countrywide without focusing on a specific area as in the active and forced migration phases.

Makedonski Telekom Example for "Launch Communication"

Makedonski Telekom mainly used media (TV, radio, print) to support the communication. The communicated messages were based on an editorial that was released with the official PR event and was supported by press releases and video to inform the market and raise awareness for the topic. Hereafter, communication material used by Macedonia during the natural migration phase is presented.

1. ATL/PR: Press release

- Macedonia to be the first Southeast European country with the new IP-based network.
- Explanation of when the modernization will start, how it will be managed, and which regions it will be comprised of at first (Figure 6.3).

2. ATL/PR: Video

- History of telephony with a relation to family, with an emphasis on different technologies used throughout the decades.

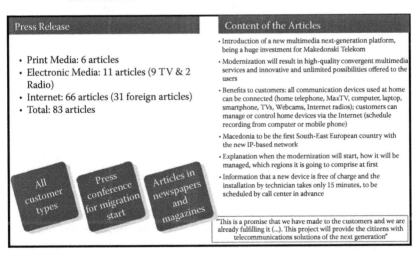

Figure 6.3 Press release of Makedonski Telekom at launch.

3. Press release was carried by media (TV, radio, online) with editorial news

- TV: Announcement of PSTN migration in the news on the main TV channels of Macedonia, private and public; news was presented on seven TV channels on the same day; newscaster talked about the start of migration highlighting additional services available to Makedonski Telekom customers.
- Radio: Announcement of PSTN migration in the news on the national radio; newscaster talked about the start of migration highlighting additional services available to Makedonski Telekom customers.
- Online: Main media placed the information online.

Telekom Deutschland Example of Natural Migration

Within the natural migration phase there is no special communication material used, except the product campaigns. Customers are approached throughout the year with DP and TP products/services for up- and cross-sell and prolongation. For a successful natural migration rate it is necessary to stop PSTN sales (or at least focus on IP) and use those customer contacts to migrate them smoothly. Communication of IP is not necessary in that context; however, it can be used to push customers to actively ask for IP products as it was done in Germany. In this case, Telekom Deutschland has focused on the comparison between the different networks, showing the quantity of features available with the IP-based products. Presenting the higher number of characteristics on IP for the same price of the standard (analogue) product should increase the awareness and interest of the consumer.

ACTIVE MIGRATION PHASE

Once the market has been educated about the new technology, countries should focus on specific areas according to the migration plan and the prioritization sequence prior to conducting and promoting trigger material to push the customer to action. Examples of communication used in Croatia and Slovakia are hereafter described.

Hrvatski Telekom Example of Active Migration

Croatia used "call-to-action" material for migration. Auto-migration and e-migration manuals are important tools to migrate customers in a cost-effective way.

The automigration is managed in a closed eco system, so-called walled garden, and the communication takes place in two stages: offline (letter), which informs customers about the migration and online (platform), which guides customers through the migration process.

Step 1—BTL: Letter for automigration was also used on the Web pages.

- Explanation that migration to the new technology is needed to ensure better services—new generation services
- Information that migration to IMS is free of charge
- List of the main benefits: faster activation of additional services, shorter period needed for installation of broadband services (e.g., IP TV), remote access in case of upgrading certain services, etc.
- Explanation of what CPE is needed for migration to IMS
- Explanation that the current tariff, tariff options, and all other features that the customer uses remain the same
- Precise instructions for PSTN migration

Step 2—BTL: E-migration (walled garden)

- Explanation that migration to the new technology is needed to ensure better services—new generation services
- Information that the customer fulfills requirements for PSTN migration (appropriate modem, etc.)
- Instructions to enter phone number and click "migrate" button
- Announcement that migration was completed successfully
- Thanks for their cooperation

Slovak Telekom Example of Active Migration

In addition to Croatia, Slovakia used the remote active migration to increase the rate of automigration. The key message used by Slovak Telekom was: "Be one of the first customers with the fixed line on a new IP technology and get a wireless desktop phone for only 1€ [$1.30] without commitment."

FORCED MIGRATION PHASE

Hrvatski Telekom Example

In the forced migration phase, Hrvatski Telekom informed its customers about the shutdown of the network and tried to motivate them to manage the migration by themselves. The communication was sent out in a form letter with an enclosed informative leaflet, differentiated according to the type of package the customer had.

6.1.2.1.2 Business Customers   Business customers require individual solutions specific to their needs; therefore some countries decided to approach their customers in person in order to define the best possible solution for the migration providing services onto the IP platform. To support the process of negotiation and be able to identify the most suitable migration plan, the key account manager (KAM)/account manager (AM), assigned as main points of contact for handling business customers, used a kind of a checklist to be filled in during each visit at the customer premises.

Such a checklist enabled a quick collection of all the essential information on the type of equipment and services that the customer had been using, and based on the analysis of customers' needs, a plan on how the customer should be migrated to an IP-based connection was defined.

Countries conducting the migration have focused on personal communication with the business customers who require individual solutions. For this reason, communication was managed by scheduling separate meetings with every business customer, handled by the KAM/AM and a technician for technical support. Countries differentiated their communication depending on the size of the enterprise. Thus, while large companies are approached by the KAM/AM, the communication with the VSE is the same as with residential customers.

Nevertheless, certain communication tools should also be available in the business area. The AM needs marketing materials such as customer presentations in order to start the discussion with his clients.

The marketing department of T-Systems in Germany, for example, has created a vast fund of materials for their salespeople. Starting with the most obvious, a high-level presentation to explain to the customers the benefits of a new all-IP environment, they also went more in depth and provided special use cases and talking points. In order to provide an insight into the vision of Deutsche Telekom to prove all-IP expertise, T-Systems also published a white paper and further communication materials.

In the beginning, they explained the role that NGN will play in the future and how they see the future of classical PSTN and ISDN. Afterward, the white paper concentrates on solution scenarios that can be built on top of an all-IP environment with the focus points on Cloud, UCC, Enterprise Mobility, and HD Voice. In the end it takes a look into the future, talking about IPv6 and new security concepts.

By sharing this paper with its customers and analysts, T-Systems proves that it understands the future trends in the network area and provides good reasons why the move to all-IP is not only unavoidable but also beneficial.

6.1.2.2 Internal Communication During migration, it is important that all employees of the operator, despite their occupation, are aware of the project, and simultaneously that they are convinced that the transformation has a positive impact for the organization itself and for their customers. Employees with a positive attitude will cause a mirror effect through their contacts with the customers, and thus will enhance the migration process.

The intensity and attractiveness of this kind of communication is highly dependent on the budget; however, even the low-cost actions can significantly raise the awareness of staff and positively impact the attitude toward the entire project.

Countries doing the migration put the emphasis on online communication in forms of: articles introducing and explaining the concept of the IP telephony, interviews with project managers, slideshows of the migration process, or informative videos explaining the IP-based packages, spreading communication on an ongoing basis from the beginning of the project kickoff. Next to online materials, some events, for example, celebration of the one-millionth customer, were planned and additionally communicated on the Intranet.

Moreover, countries developed tools to support the salespeople on how to deal with customers. For example, a manual explaining the most important points to stress with customers as well as the hardware and the installation procedure was created.

KEY LEARNING

- Communication materials must be differentiated according to the customers and migration phases.
- Take into account the time needed for the production of some materials and include it in the planning.
- Communicate clear messages and keep promises.
- An official letter for supported migration with technicians had a high effectiveness (response rate 80–90%) in some countries.
- Business customers need to be approached by AM/KAM to identify the usage of the network with additional services.
- Internal communication to the complete company provides a further use of the employees as multiplicators to the market.

6.2 Sales Channel Plan

Within the PSTN migration it is critical to migrate customers as soon and as smoothly as possible. Therefore, each potential customer contact should be treated as an opportunity to migrate him to the new technology through natural migration. At the same time, it has to be made sure to not lose any potential for up-/cross-sell once approaching the customer actively.

6.2.1 Steering and Orchestrating the Sales Channels

The decision on the sales channels used for customer contact should be based on the defined migration plan, the availability of specific sales channels in each country and prioritized areas, as well as lessons learned coming from the prioritized regions.

6.2.1.1 Preferred Sales Channels as Customer Touch Points

6.2.1.1.1 Residential Customers and Very Small Enterprises (VSEs) For each of the migration phases, whether natural, active, or forced, different sales channels are more or less suitable for initiating the migration. Thus, it is crucial to identify the most preferable sales channels for the migration, taking into account each customer segment separately (residential, VSE, or business).

For residential and very small enterprise (VSE) customers, all sales channels should be considered as potential customer touch points, existing in each of the phases. Therefore, all sales channels need to be trained to at least explain the main benefits.

A prerequisite for the highest success in natural migration is the stop of PSTN sales, which allows using every customer contact to do the migration. In case it is not possible to stop PSTN sales for all customer segments at once, a stepwise approach can be chosen:

0. Stop of PSTN sales for TP (with VOBB capable routers).
1. Stop of PSTN sales for DP.
2. Stop of PSTN sales for business customers (natural migration should start as early as possible as it will take longest).
3. Stop of PSTN sales for other customer segments.

Natural migration begins with the official start of the PSTN migration (with mass products). Customers initiate the contact with the operator themselves, all sales channels available within the NatCo/region should be ready to migrate the customer and stay involved to use the full potential of migration.

In the active phase, the main contact channels are telesales to manage outbound sales calls and technicians who clear faults or help in supported migration. The key contact channels, aligned with the migration strategy, will play a major role in the success of migration. Simultaneously, other sales channels should still serve as a support to the preferred channels to avoid untapped migration opportunities. During this phase, providers will have to deal with sleeping lines, which may represent a risk and revenue loss not only for business customers, but even for residential customers.

In the forced phase of migration the key contact channel is represented by technicians.

6.2.1.1.2 Business Customers   For medium/large business customers, the only sales channel is the key account manager (KAM)/account manager (AC) who is able to handle the relationship with their customers, which often requires a different approach than with the residential or VSE segment. The role of the (key) account manager is essential when contacting especially large enterprises; as such cooperation is run on special rules, depending on the relationship with the customer and often related to solutions needed. Sometimes it is crucial that in the KA/LA segment the account manager is accompanied by a technician when presenting the migration plan to the customer. This has to be taken into account during the resource planning.

6.2.1.1.2.1 Specific Challenges for KA/LA Customers   KA/LA customers usually have specific environments customized to their needs. Every customer will bring his own challenges during the migration, therefore a separate evaluation is needed for each one. For these customers there will be no standardized approach, just the same as in the regular business (Table 6.2).

Table 6.2   Specific Challenges for Key Account (KA) and Large Account (LA) Customers

	CHALLENGE	SOLUTION
High complexity	The technical landscape of large business customers is usually highly complex. In order to guarantee a smooth transformation every line has to be analyzed individually to find the right replacement product.	Work with the largest customers from the early beginning, without underestimating the needed effort. Only the analysis and the migration plan for LA can take more than 1 year, therefore employing a large amount of resources. Including the negotiation process it can be a 2-year process for only one customer.
Customer with sleeping lines	Numerous customers have "sleeping lines," meaning they have phone or data lines with no traffic on them. When approached for migration, these customers will take a critical look at their infrastructure. In most of the cases this will lead to revenue loss because of cancellations. This is not a challenge specific to the IP transformation project. The risk that a customer does an audit of his infrastructure and discovers sleeping lines is always imminent.	There is no silver bullet for this challenge: customers will cancel as soon as possible lines they do not need. To avoid revenue loss, upsell to higher value services during the migration. It is a chance to increase the trust and standing with the customer by telling him that, during the redesign of his environment, we found several lines he does not use. This might lead to a slight decrease in the recurring revenues. However, it is a clean cut and, after the migration process, there will be no more hidden risk in the fixed-line revenues.

Table 6.2 (*Continued*)   Specific Challenges for Key Account (KA) and Large Account (LA) Customers

	CHALLENGE	SOLUTION
Customer is not willing to change	There are several reasons behind this. He might fear downtime and/or that the new products are not as reliable as the ones he is used to ("never change a running system").	Use workarounds so that he can be quietly migrated. Only possible for customers with basic services. Customers with complex environments have to be convinced. For that there are two basic levers: • Best practice and references: skeptical customers might change their mind by seeing the benefits of the new IP services and testimonies of smooth transformation. • Financial benefits: When a customer is too important to lose and is still not convinced, put together for him a special package with higher value services for a very attractive price.
Not all ISDN products have good replacements in the IP world	There are customer environments that cannot be migrated 1:1 to the IP world. Customers will be even more skeptical to change and experiment with their working solution.	For these customers a complete redesign of their environment is necessary. A lever for that can be a package deal with some add-on services for free. If a customer cannot be convinced interim workaround solutions have to be found.
Unwillingness of business customers to change CPEs	Some customers might not be willing to change their customer premise equipment because it has not been written off, for example.	Offer to trade in the equipment that has to be replaced for a symbolic price. For the most important customers offer to buy back the old equipment and give the new CPEs a leasing or renting model: transforming customer CAPEX to OPEX.
Multinational companies	Some very large customers have strong requirements from their headquarters to use only ISDN-based PBXs in their locations.	Almost all of these customers on a global level are T-Systems customers. Therefore an alignment with the global account manager is necessary to work out a solution with him. One NatCo could act as a pilot as sooner or later all the networks globally will change to IP.

6.2.1.1.2.2 *Commercial Risks in the B2B Area*   Taking into consideration the particular nature of the KA/LA customers and their related challenges, it was clear from the beginning that the IP transformation would include several commercial risks. In addition to the specific challenges illustrated above, further threats on both the cost and the revenue side should be considered.

On the cost side, the migration process in the KA/LA segment is strongly resource intensive since every line and product of the customer have to be evaluated individually for what is used to find the right IP-based replacement.

On the revenue side, for example, IP is still considered in most of the countries as a cheaper product. Therefore, the market demands discounts on comparable products.

Consequently, all in all, the fixed voice revenues can significantly drop in the B2B segment. In order to minimize these risks as much as possible, a push to cross-/upsell IP/cloud-based products is necessary. Examples for that are IP PBX systems either hosted or on-site. An IP Centrex solution can address most of the midmarket and can be bundled with the new VoBB products.

KEY LEARNING

- Use every customer contact point to migrate customers. Although technicians are not strictly a sales channel, they hold a potential to become a sales point in the contact with customers, as the customer trusts them.
- Technicians should be used as salesmen or lead generators, but there should be a developed mechanism for quality check.
- Check and analyze customer databases at the beginning of the project to identify customer needs and to prepare a strategy for cross- and upsell.

6.2.1.2 *Channel Focus per Migration Phase*   The effectiveness of PSTN migration strategy requires proper steering and orchestration of different sales channels, which thus increases successful alignment of communication and sales strategies.

Hence, based on the migration plan, which includes the plan for product development sequence and sequence of targeted areas, the sales strategy is being determined (see Figure 6.4).

Supervised by the central organization, local teams from each of the targeted regions, including different functional areas such as sales, promotion, and technical service, work closely together to define a strategy suitable for their specific region taking into account different customer profiles. The involvement of the central team assures that the PSTN sales strategy is aligned with the overall corporate strategy and respectively assures proper channel steering.

Depending on the migration phase (natural/active/forced), different sales channels are involved; therefore, it is crucial to assure that their work is orchestrated for better planning and effective execution of sales strategy. As a result, each of the targeted teams will be able to develop a campaign specific for their region and customer type coordinated by the regional manager to steer the sales channels and to coordinate contact with the customer.

Orchestration in respect to PSTN migration means assigning different roles for each particular channel in the sales process within different migration phases and assuring that all sales channels work together hand in hand not to lose the customer in the sales funnel. Orchestration takes place within both push and pull channels; it is important to identify the decision factors for when to use what kind of sales channel.

Steering of the channels requires some basic information, such as geo marketing results, for example, for evaluation of areas for the migration, CRM customer database with an overview on existing products, and so on. Based on this input the evaluation of optimal sales channel mix is conducted, allowing further steering of channels to give a clear guidance to all sales channels when to approach which customer type with what kind of product.

The orchestrated channels need to be further monitored and evaluated. This allows the review of the sales channel plan and its adjustments, accordingly. The redeveloped strategy serves as a base for further channel steering. The customer database is used to identify customer needs and fully use upsell/cross-sell potential. The process of steering can only be effective if strong customer database (CRM) and geo marketing tools are in place.

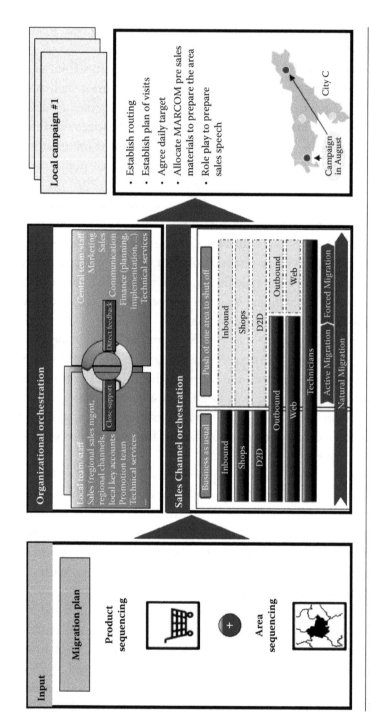

Figure 6.4 Development of local campaigns based on steering of sales channels.

6.2.1.2.1 Early Push of the Natural Migration Countrywide Steering all the sales channels from the start of the migration allows an effective management of all the possible customer contact points. A push in the natural migration brings a twofold advantage to the company. On one hand, the cumbersome process of the network switch is shorter and safer (lower churn) since the customer is not disturbed in additional contexts. On the other hand, the costs sustained by the firm for the migration will be lower in this first phase.

From a company's perspective, one of the reasons of conducting PSTN migration is the high cost related to maintenance of the old platform as well as its energy consumption. Additionally, as migration to IP requires huge investments, the cost of migration needs to be taken into consideration as the different phases generate costs necessary for facing the customer (iAD, sales and communication costs). Companies should carefully assess the cost structure in order to evaluate the best migration strategy to pursue.

In order to minimize the migration costs, NatCos successfully focused on the natural migration from the early start of migration, allowing them to achieve significant rates during the migration period.

Two key prerequisites have driven this success:

1. Availability of substitute products (mirroring old and IP portfolio) from migration start.
2. Establishment of smooth processes supporting and focusing on IP.

The first aspect refers to the early development of several IP products in the portfolio, which mapped all the features available in the PSTN portfolio. When starting the migration, it was not necessary to develop new products and offers specifically for IP as the countries have a technology neutral strategy. A market research conducted in July 2013 has revealed that customers migrated to IP did not notice any major problem with their network, therefore proving the quality of the IP products. Moreover, it is important that special cases and features are identified from the beginning in order to elaborate a workflow process after.

The second condition of this success is related to the smooth processes enforcing IP sales, both at a technological and an agent level. During all customers' contacts in stores, the IT-system automatically presents to the sales agent only the IP products. In case the agent identifies one of the special cases and the old legacy product is needed, this workflow has to be manually run and monitored. As demonstrated from the high rate of natural migration reached, customers are smoothly and successfully migrated, despite the hurdle of signing a new contract.

KEY LEARNING

- The later the stage of the migration, the higher the cost associated with it. Cost structure differs depending on the type of service the migrating customer has (broadband or voice only), where for voice only the cost is much higher as it requires an installation of new equipment.
- Cost in the natural phase of migration is still relatively low; therefore, the migration should be pushed in this phase, even if it would require additional incentives to the agents or the customer.
- Prerequisites before migration start: Substitute IP products and supporting processes to be carefully developed and implemented in order to successfully push the rate of natural migration and limit the customers' churn due to migration.
- Start natural migration as early as possible in all segments (residential and business).

6.2.1.2.2 Local Push in Active and Forced Migration to Shutdown Areas
The successful targeted PSTN migration strategy requires alignment of all sales and communication channels. Whereas communication within the natural migration phase takes place in a continuous and countrywide level, further phases require the definition of local push and follow-up activities.

The End-to-End process is used to optimize the process of migration and to increase its efficiency. The process is managed in four steps with upfront planning and preparation of the local campaigns. Upfront communication in the natural migration phase is necessary to be managed before starting local push for cost reasons, where all the communication plays a more informative role. It is recommended to conduct active migration with strong local push within the targeted areas to manage as many migrating customers as possible and to avoid or at least minimize the need of forced migration. In order to maximize the efficiency of the migration process, local teams have to be closely aligned with the central organization.

For a strong local campaign, a special product/proposition is necessary with regulatory approval. Once the local push team is set up and it identifies the targeted customers, upfront communication will prepare the local region. Then the local push can start with identified sales and communication channels.

Despite a strong local push, not all customers can always be migrated during the active phase. Due to increasing cost, the remaining customers should be switched as soon as possible, which is after all the attempts to contact or to migrate the customers turned out to be unsuccessful. Cleaning the areas should be managed as smoothly as possible; therefore, it is necessary to develop a specific process and relative timeline for better planning of the shutdown operation, adapted accordingly with the type of customer and type of service at customer premises. Depending on country-specific requirements, forced migration should be considered up to 30 days before an area will be switched off. The communication to shut down an area has to be initiated further in advance depending on the regulator.

KEY LEARNING

- Forced migration is in some cases the only migration solution left; however, due to booming cost it should be avoided. Thus, additional incentives for the sales force to increase the migration rate and to further motivate them will still be at a lower cost than it would by relying on a stronger forced migration.
- As in the PSTN migration project, different sales channels need to work hand in hand; it is required to establish a new commission system in the organization.

6.2.1.3 Reporting To support further definition of the migration strategy, the process of PSTN migration has to be supervised and controlled. Each operator should define a set of KPIs to consistently monitor the usage and success rate of sales channels within the migration. The monitoring is necessary to keep track of the migration process and define potential critical points that must be solved immediately (Figure 6.5).

The efficient reporting structure allows monitoring of the status of PSTN migration and feeds the database as a source for channel steering. It is necessary to plan the reporting structure on time, before going into the migration process.

Sales migration reports must enable effective monitoring and management of sales channels in the migration process.

Similar to the reporting structure in the case of PSTN migration, a reporting structure for tracking of cross- and upsell processes should be introduced, with the possibility to view by products and sales channels during the migration process. The reporting structure for tracking realization of retention offers, by products and sales channels during the process of migration, should be implemented within the reporting system for further review of the sales approach and strategy redefinition.

KEY LEARNING

- A reporting structure should be developed in order to be able to track the flow of migration and sales results, and thus adjust the strategy if needed.
- The new reporting structure must be developed on time, which is before the PSTN migration process, as the collected data is essential for steering and strategy definition. It may require adaptation of IT systems.

6.2.2 Cross-Sell, Upsell, and Retention

Once a customer is approached for PSTN migration, leverage of the full potential to generate higher revenue and retention needs to be ensured.

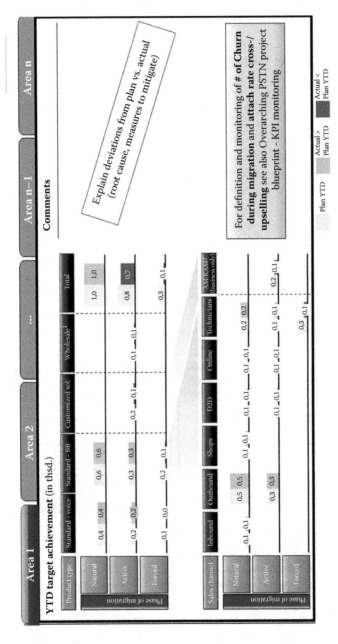

Figure 6.5 Dashboard—sales channel monitoring per product type and sales channel.

The sales plan should include a list of defined up-/cross-sell and retention moments aligned with overall sales strategy and migration plan. Therefore, definition of customer touch points for cross-/upsell and retention is also structured along sales channels and migration phases.

Makedonski Telekom Example

Cross-/upsell activities can be undertaken even after the PSTN migration has been conducted, thanks to the introduction of new IP products. As a result of the all-IP strategy followed, new services can be designed for pure voice customers. In this respect, Makedonski Telekom has launched *Broadband on-demand*.

Broadband on demand (launched on June 20, 2013) is intended for occasional Internet users (i.e., weekend house owners), who can immediately activate the service through IVR without an additional CPE needed. The service is sold in three versions: 24 hours, 3-day, and 7-day package.

On the other side, another offering, Try & Buy (launched in October 2013) aims at lowering the barrier for cross-selling, by offering to nonbroadband customers a 1-month test of the broadband Internet service, without any further obligations (Figure 6.6).

These offerings are only feasible with a full VoBB strategy and not with MSAN POTS and solutions.

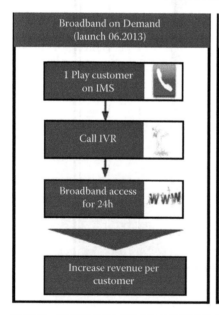
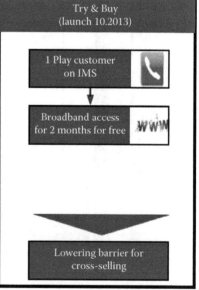

Figure 6.6 New IP-based offerings in Makedonski Telekom.

KEY LEARNING

- Pure voice customers are hard to migrate to VOBB and even more difficult to do as an up-/cross-sell because they are price sensitive. However, thanks to the all-IP strategy, new offerings for pure voice customers can be introduced, increasing the up-/cross-sell potential.
- In general, cross-/upsell can be leveraged by focusing on the additional services and migrating the customer to the IMS platform.

6.2.2.1 Retention   Similar to cross-/upsell, the most suitable retention moments must be identified for each of the migration phases and customer types. Country examples show that within the active migration phase the opportunity for retention is highest, especially by focusing on:

- *Automigration*: The potential for retention in combination with a leaflet; in order to increase retention, it is possible to combine the migration notification with potential retention offers that can be ordered and by communicating to the customer sales channels where those offers may be ordered (e.g., online, via telesales/outbound, at the shop).
- *Technician*: The ability to sell possible retention offers (sales/sales lead) during the process of PSTN migration. In order to increase the success, these offers should be personalized to the customer and activated immediately upon migration.

KEY LEARNING

- Double-check regulatory constraints—do not bring yourself to a position where you must stop an offer in the middle of the project.
- Prepare IT systems before starting migration to receive relevant reporting.
- Incentivize the employees and keep an eye on budget spending.

In line with the focus on natural migration, countries are successfully exploiting the retention and prolongation moments to migrate the customers to VoIP, without having to allocate additional resources for it.

6.3 Training

Migration as a new activity requires certain additional skills from all functional groups involved in the process. What is more, it requires development of skills suitable for dealing with both residential and business customers. Moreover, due to the importance of the project inside the company, an all-employee training, focusing on the main aspects of the IP transformation and its benefits, should be conducted. This requirement originates as well from the cross-functional set up of the project, which involves a close interaction among the different work streams.

An organized approach for training is introduced in Figure 6.7.

Training for each functional group differs from each other, as each group plays different roles within the migration and during contact with the customers.

Sub-activity	Challenges of PSTN migration
Customer Care	• Before migration: migrate customers that call in for any other reason (how to lead the call) • During migration: support customer in migration process; get to know new processes + new services • After migration support: need to identify the issue in case customer calls in
Technology	• Before migration: 2-in-1 migration to be trained + generated lead for x-/upsell • During migration: support customer in migration process; get to know new processes + new services • After migration support: managing customer complaints
Sales Force	• How to overcome the negative customer perception (benefit story) • Make sure all sales channels are trained or at least informed about key steps of migration
External Company *depending on function in the migration process	• Keep Code of Conduct • Generate leads and/or x-upsell • Get to know new processes + new services
KAM/AM	• Prepare a training content adjusted with the needs to handle business customers (KAM, AM)
All-Employee	• Make sure all employees are aware of the keypoints of the PSTN migration project

Figure 6.7 Functional groups for training.

Thus, training for involved groups will have a different scope and focus as it requires varied types of skills. Within the training structure, four main training needs were identified:

- *Product/proposition*—To familiarize with products/offers for the best migration expertise and fault resolution, as well as to assure migration efficiency and use of cross-/upsell potential.
- *Sales training*—To assure that functional groups have best sales techniques to be able to recognize customers' needs and generate up-/cross-sell potential.
- *Fault clearance*—To serve the best customer experience by immediate fault resolution and customer guidance in order to increase the effectiveness of migration, and at the same time to recognize customers' needs and generate cross-/upsell potential.
- *Front-end application/processes*—To familiarize with the new approach and any potential changes in the processes related to new services.

The entire training for PSTN migration has to be proceeded by joint preparation of the IT systems, application processes for new services, and the final offer. Parallel to the preparation of the training content, it is important to define the provisioning process at the customers' premises, as well as to do a test run of the application to avoid any inaccuracies in the guidelines given in training, as it can affect the effectiveness of the PSTN migration process. Only after all the issues are clarified, the training for all functional groups can take place. The training phase is assumed to last 3 months in total and to be completed before the official start of the migration.

Migration as a new activity requires certain additional skills from all functional groups involved in the process. Training must be prepared specifically for each of the functional groups, with adjusted content depending on needs and objectives set. Therefore, it is important to keep in mind that there will need to be at least 3 months for all training to be done. Moreover, for a successful migration it is necessary to have a test run upfront with all necessary front-end application running and to make sure that all zero processes are established.

6.3.1 Sales Force Training

One country has built a digital training for the sales force composed of the following elements:

- e-guide
- e-training in online academy
- Short video with tips
- Training in the form of a detailed presentation

Such diversification of education ensures that the participant is both entertained and thoroughly prepared for selling the new products and services of IP telephony.

Digital sales training for residential customers can be regarded as a best practice for digital as it explains the IP platform through a "step-by-step" or "for dummies" approach with an attractive layout by simply using functions of available basic software.

First, digital training is differentiated from content perspective, so that it increases the attractiveness for the participants and keeps their attention. It covers different aspects that are essential for the successful sale of the IP packages, but at the same time it does not overload the participants with information. Topics, such as technological development or tariff changes, are presented in relation to the changes that have occurred within the organization, thus it helps salespeople to identify themselves in the project. Simplified but specific tariff overview and guidance on the upsell options allows for quick acquisition of the information. In addition, there is a dedicated part for recollection of basic selling techniques, which was extended for dealing with more sophisticated customers. The sales force is also prepared from the cabling side, to be able to immediately provide guidance to the customers in case of technical questions. Most of the content, and especially processes, are visualized in a very uncomplicated and clear way, in the form of a table of process flow. Another positive tool is to include a quiz in the e-training lever, where the participant is asked to match the right answers, which allows practicing and memorizing the content on spot.

6.3.2 Key Account Manager (KAM) Training

Training modules and content differ not only depending on the functional group that needs to be trained, but also according to customer profile (e.g., residential, business). In order to deal with business customers, who require a different approach than residential, countries prepared training programs for their sales departments, both direct and indirect. Training programs were conducted in the form of a presentation with the emphasis on examples from migration practice and interactive discussion on the subject of migration process. Business training was built specifically to cover peculiar issues (e.g., processes or solutions not in place) in the country.

The business training roadmap should always be adjusted according to the needs of each country. For instance, the number of rounds for F2F training can vary depending on regional needs (number of KAM too high to handle in one group; region too huge to gather all participants in one place). At the beginning of the migration it is advisable to offer assistance (coaching) in field activities as part of training. This increases the level of expertise in dealing with such sensitive customers as those in the business sector. However, it should be taken into account that both the increased number or rounds, as well as on-site coaching, may require increased resources.

6.3.3 General "All-Employee" Training

Due to the high importance of the PSTN migration project and the responsibility that the organization bears in managing the migration as soon as possible, it is necessary to build awareness and educate all employees to ensure that all opportunities for the migration are used. Thus, it is advisable to introduce another training session for all employees. Such training can be easily prepared at a low cost without the use of professional tools.

KEY LEARNING

- Full transparency of changes due to PSTN migration.
- Dedicated cross-functional teams to follow/solve key issues.
- Refresh sales/CC staff knowledge on a regular basis.
- Assure enough time for preparation of proper training.

6.4 Migration Support Customer Care and Technical Service

One challenge in the execution of PSTN migration is the optimization of required resources from customer care and technical service during and after the migration of customers. Indeed, resources are needed not only for the process itself, but also for faults derived from the migration.

In detail, this section explains how customer care and technical service should be involved in a PSTN migration project and how demand volume should be planned.

6.4.1 Involvement of Customer Care and Technical Service in the PSTN Migration Process

Effective involvement of customer care and technical service in the PSTN migration project is not only about availability of resources; it is also about having the IT systems ready in order to track call reasons during the migration process.

With regard to system readiness, it is strongly recommended to check the need for system modifications during the conception phase of PSTN migration to secure on-time readiness of IT systems and a sufficient timeframe for training of employees.

A knowledge base and regular tracking of call reasons should be implemented before the migration starts. It has to be emphasized again that PSTN migration is not about finding an 80/20 solution; this means that every customer has to be migrated from the old legacy system to the new IP world. If an operator lacks the possibility of tracking call reasons, it will be difficult to optimize utilization of migration types and to discover issues with customer experience during migration. Figure 6.8 provides an overview of possible call reasons related to PSTN migration and new IP technology. Calls can happen when migration support is needed, due to faults caused by the migration within a given timeframe, and general faults because of IP technology (long term). Furthermore, one should consider explicitly tracking cancellation calls along the migration process.

From the early beginning of the PSTN migration project, the need for customer care and technical service support for migration has to

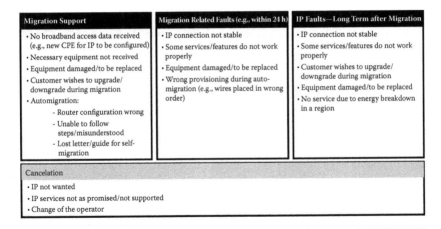

Migration Support	Migration Related Faults (e.g., within 24 h)	IP Faults—Long Term after Migration

• No broadband access data received (e.g., new CPE for IP to be configured)
• Necessary equipment not received
• Equipment damaged/to be replaced
• Customer wishes to upgrade/downgrade during migration
• Automigration:
  - Router configuration wrong
  - Unable to follow steps/misunderstood
  - Lost letter/guide for self-migration
	• IP connection not stable
	• Some services/features do not work properly
	• Equipment damaged/to be replaced
	• Wrong provisioning during auto-migration (e.g., wires placed in wrong order)
		• IP connection not stable
		• Some services/features do not work properly
		• Customer wishes to upgrade/downgrade during migration
		• Equipment damaged/to be replaced
		• No service due to energy breakdown in a region

Cancelation
• IP not wanted • IP services not as promised/not supported • Change of the operator

Figure 6.8 Call reasons related to IP.

be planned to secure availability of resources on time and in expected quality. Therefore, resource planning needs to be strongly linked to overall migration strategy, meaning the migration plan (e.g., product and area sequencing) and the underlying assumptions on application of different migration types (e.g., automigration, supported migration, 2-in-1 migration).

Assumptions on resource needs and actual demand should be constantly tracked and monitored during migration to ensure the quality of the process (customer perception and response rate in case of faults) and achievement of migration targets. In the next section, the approach for planning resource demand will be explained in more detail.

KEY LEARNING

- Track call reasons during and after PSTN migration.
- An automated testing procedure secures synchronization of customer care with other functions (e.g., technicians).

6.4.2 Planning of Demand Volume

The incremental change in demand volume due to PSTN migration should be planned for migration support, related faults, and long-term impact of IP versus PSTN (see Figure 6.9).

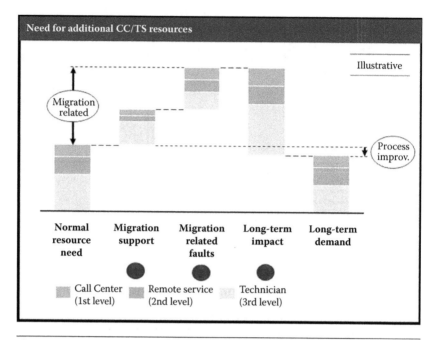

Figure 6.9   Resource planning approach.

In detail, total demand volume for call center agents (first-level support), remote service (second level), and technicians (third level) will be influenced by the following factors:

1. *Migration support*: Mix of migration types per country determines the need for resources per customer segment/product or feature and period. Availability of internal and external resources determines total capacity and thus maximum migration volume. From experience, an additional amount of resources is needed during this stage. Therefore, a high focus should be placed on the most cost-efficient activities, auto-migration and natural migration, in order to restrain where possible the amount of resources.

2. *Migration related faults within a given timeline*: It is crucial to make assumptions and plan additional resources for handling faults due to the migration process (e.g., wrong provisioning of service, problems with IMS connection). In peak phases of migration this category can easily cause a high demand for resources and, therefore have a significant impact on the cost of the project. It is especially important

in the early project phase to plan for a higher demand as the organization and involved functions are still in a learning circle. Together with call reason analysis, the actual demand from faults after migration should be used to optimize migration in the first step.

3. *Long-term impact*: The number of IMS faults and, therefore, of inbound fault calls is expected to increase not only in the 24 hours following the migration but also in a longer period. As of today, there is no valid data on long-term impact on the long-term migration impact, since operators are still migrating the entire customer base or have recently concluded. On the one hand, a shorter mean time between failures is expected to cause a higher call volume, but on the other hand, process improvements and the higher stability of the IP services may reduce long-term demand.

In general, the planning for demand volume should be made on a daily basis to ensure resource availability on each support level, closely aligned with the migration plan. This includes expected migrations from natural migration as well as active and forced migration for focus areas.

KEY LEARNING

- The challenge for migration support is to optimize need for direct customer contacts—to reduce customer care calls, but especially technical service visits.
- Expect a significant call and fault volume after the migration, especially in the early phase of the project. Additional resources have to be planned carefully to secure customer satisfaction.
- A call-tracking system can ensure a faster response and a better understanding of the faults and, consequently, lower the migration costs.
- First empirical data show a higher call volume for customers on IMS versus PSTN.
- So far there is no long-term experience available on resource demand change after a successful PSTN migration.

It is strongly recommended to increase the migration volume slowly in the early stage of the project, for example, by first pilot areas and natural migration start. Therefore, it will be possible to check the stability of migration processes, to enable sufficient on-job training of employees in all service levels (learning curve), and to validate assumptions on resource need per migration.

Index